The Agricultural Revolution

About the Editors

Dr. Traloki Singh is working as a Subject Matter Specialist (Agronomy), ICAR-CSSRI Krishi Vigyan Kendra Hardoi-II, Sandila, Hardoi-241203 (U.P.). He did B.Sc. (Agriculture) from Dr. B. R. Ambedkar, University, Agra in 2002 and M.Sc. in Agronomy from Dr. B. R. Ambedkar, University, Agra in 2004 (R.B.S. Collage, Bichpuri, Agra) (U.P.) and Ph.D. in Agronomy from Bundelkhand University, Jhansi in 2010 (U.P.). He has also Qualified ASRB-NET in 2010. Dr Singh has more than 15 years of experience in the field of Research & Extension in various organizations of ICAR (CAZRI, CSSRI, ATARI, and NAIP Project). He has various research paper, book chapter, abstract, and article and book co-author in national and international publications. He has received Young Extension Worker Award and Best Scientist Award in the year 2017 and 2018.

Dr Somendra Nath is working as a Subject Matter Specialist (Agronomy), Krishi Vigyan Kendra Ballia (U.P.). He did B.Sc. (Agriculture) Hons from Janta Maha Vidhyalaya Ajeetmal, Auraiya in 2003 and M.Sc. in Agronomy from Chandra Shekhar Azad University of Agricultural and Technology, Kanpur in 2006 and Ph.D. in Agronomy from Acharya Narendra Deva University of Agriculture and Technology, Faizabad (Ayodhya) (U.P.). Dr Nath has 13 years of experience in the field of Research & Extension. He has various research paper, book chapter, abstract, and article in national and international publications.

Mandeep Kumar is currently pursuing Ph.D. in Agronomy from Chandra Shekhar Azad University of Agriculture and Technology, Kanpur. He has accomplished BSc. (Ag.) and M.Sc. in Agronomy from Chandra Shekhar Azad University of Agriculture and Technology, Kanpur (U.P.). Mr. Kumar has got many awards like Best M.Sc. Thesis Award, Young Professional Award, and Best Research Scholar Award and also has various research paper, book chapter, article, abstract and book co-author in reputed national and international journals. He has also participated in many national and international seminars and conferences.

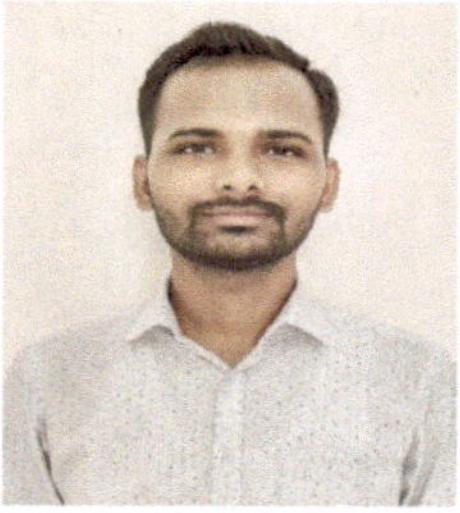

Ankit Kumar Maurya is currently pursuing Ph.D. in Agricultural Economics from Chandra Shekhar Azad University of Agriculture and Technology, Kanpur. He has accomplished BSc. (Ag.) from Udai Pratap Autonomous College, Varanasi and M.Sc. in Agriculture Economics from Banda University of Agriculture and Technology, Banda (U.P.). Mr. Ankit has got many awards and also has various publications in reputed national and international journals. He has also participated in many national and international seminars and conferences.

Dr Chander Bhan is working as Associate Professor (Horticulture) at College of Agriculture (Swami Keshwanand Rajasthan Agricultural University, Bikaner), Hanumangarh (Rajasthan). He did B.Sc. (Agriculture) Hons. and M.Sc. (Agriculture) Horticulture from Swami Keshwanand Rajasthan Agricultural University, Bikaner and Ph.D. (PHT) from ICAR-Indian Agricultural Research Institute, New Delhi. He has also qualified ICAR-ASRB National Eligibility Test (NET) in Horticulture. He has various research paper, review papers, book chapter, abstract, article and booklets/manuals in national and international publications. Dr Bhan has got many awards like District administration, Sriganganagar on 15.08.2016 for doing excellent official work, Young Scientist Award 2018, Best Oral Presentation Award, Outstanding Horticulturist Award 2020, appreciation certificate awards.

The Agricultural Revolution
A Journey Towards Sustainable Agriculture

Traloki Singh
Subject Matter Specialist (Agronomy)
ICAR-CSSRI, KVK, Hardoi-II
Sandila, Uttar Pradesh

Somendra Nath
Subject Matter Specialist (Agronomy)
KVK, Ballia, Uttar Pradesh

Mandeep Kumar
Ph.D. Scholar (Agronomy)
Chandra Shekhar Azad University of Agriculture and Technology
Kanpur, Uttar Pradesh

Ankit Kumar Maurya
Ph.D. Scholar (Agricultural Economics)
Chandra Shekhar Azad University of Agriculture and Technology
Kanpur, Uttar Pradesh

Chander Bhan
Associate Professor (Horticulture)
College of Agriculture (SKRAU)
Hanumangarh, Rajasthan

PMW

PMW

FU 75, Pitam Pura
New Delhi 110034
publishmywork23@gmail.com
Mob. 91-8447075807

Print ISBN: 978-93-61341-81-6

ebook ISBN: 978-93-61340-44-4

Preface

The urgency of revolutionizing agriculture through technology stems from the critical intersection of food security, environmental sustainability and global population growth. With the world's population projected to reach 9.7 billion by 2050, the demand for food will surge, necessitating a transformation in agricultural practices to meet this demand sustainably.

In the present, technology-driven solutions offer immediate relief to challenges such as climate change, resource scarcity and rural poverty. Precision agriculture techniques, enabled by drones, sensors, and satellite imagery, optimize resource utilization by precisely delivering water, fertilizers and pesticides, thereby reducing waste and environmental impact. Additionally, the adoption of genetically modified crops engineered for drought resistance, pest resilience, and increased yields bolster food production in the face of changing climatic conditions.

Looking towards the future, technology holds the promise of further advancements that can redefine the agricultural landscape. Vertical farming and hydroponic systems maximize land use efficiency while minimizing water consumption, offering scalable solutions for urban environments and regions with limited arable land. Furthermore, breakthroughs in synthetic biology could revolutionize crop breeding, accelerating the development of resilient and nutrient-rich varieties to address nutritional deficiencies and enhance food security. However, it's crucial that these technological advancements are deployed with a steadfast commitment to sustainability. Embracing agro ecological principles, such as diversified cropping systems and natural pest management, alongside technological innovations ensures the preservation of biodiversity and ecosystem health while fostering resilient food systems.

In essence, the tech-driven revolution in agriculture is not merely about increasing yields or profitability; it's about safeguarding our planet's future while ensuring equitable access to nutritious food for all. By harnessing the power of innovation and sustainability, we can cultivate a food system that nourishes both people and the planet, securing a brighter and more prosperous future for generations to come.

Editors

Preface

The urgency of revolutionizing agriculture through technology stems from the critical intersection of food security, environmental sustainability, and global population growth. With the world's population projected to reach 9.7 billion by 2050, the demand for food will soar, necessitating a transformation in agricultural practices to meet this need and sustain [illegible].

In the present, technology-driven solutions offer [illegible] to challenges such as climate change, resource scarcity and food insecurity. Precision agriculture techniques, enabled by drones, sensors and satellite imagery, optimizes resource utilization by precisely delineating water, fertilizers and pesticides, thereby reducing waste and environmental impact. Additionally, the adoption of genetically modified crops engineered for drought tolerance, pest resistance, and increased yields bolster food production in the face of changing climate conditions.

Looking forward, the future technology holds even greater [illegible] advancements that can redefine the agricultural landscape. Vertical farming and hydroponic systems maximize land use efficiency [illegible] water consumption, offering scalable solutions for urban environments and regions with limited arable land. Furthermore, breakthroughs in synthetic biology could revolutionize crop breeding, accelerating the development of resilient and nutrient-rich varieties to address nutritional deficiencies and enhance food security. However, it is crucial that these technological advancements are deployed with a steadfast commitment to sustainability. Integrating agroecological practices, such as diversified cropping systems and natural pest management, alongside technological innovations ensures the preservation of biodiversity and ecosystem health while fostering resilient food systems.

In essence, the tech-driven revolution in agriculture is not merely about increasing yields or productivity; it's about safeguarding our planet's future while ensuring equitable access to nutritious food for all. By harnessing the power of innovation and sustainability, we can cultivate a food system that nourishes both people and the planet, ensuring resilience, and [illegible] prosperous future for generations to come.

Editors

Contents

1

Role of Technologies to Precise Farm Management

Somendra Nath

Subject Matter Specialist (Agronomy), KVK, Ballia, Uttar Pradesh

Abstract

The importance of technology in precise farm management cannot be overstated in the contemporary agricultural landscape. As the global population burgeons, arable land diminishes and climate change poses unprecedented challenges, technology emerges as a vital tool for sustainable and efficient agricultural practices. At the forefront is precision agriculture, which harnesses innovations such as GPS, drones, and sensors to optimize resource utilization. By precisely mapping soil conditions, moisture levels and crop health, farmers can tailor their interventions, conserving water, reducing chemical inputs, and maximizing yields. Moreover, data-driven decision-making facilitated by advanced analytics enables farmers to make informed choices in real-time. By analyzing vast datasets encompassing weather patterns, market trends and agronomic insights, farmers can mitigate risks, enhance productivity and bolster profitability. Furthermore, automation and robotics revolutionize traditional farming operations, minimizing labor costs and increasing efficiency. Automated machinery for planting, harvesting, and monitoring accelerates processes while ensuring accuracy and consistency. In essence, technology serves as the linchpin for sustainable and precise farm management, enabling farmers to navigate the complexities of modern agriculture while fostering resilience and productivity in the face of evolving challenges.

Keywords: *Technology, Optimum Resource Utilization, GPS, Sustainable Practices, etc.*

Introduction

In the vast tapestry of challenges confronting humanity, few are as fundamental as ensuring a secure and sustainable food supply for a burgeoning global population. As the world's inhabitants surpass 9 billion by 2050, the pressure intensifies on agriculture to deliver higher yields, reduce environmental impact and adapt to the uncertainties of climate change. In this pivotal moment, the convergence of technology and agriculture emerges as a beacon of hope, offering unprecedented opportunities to revolutionize farm management and secure our agricultural future. The role of technology in precise farm management has transcended mere convenience to become an indispensable cornerstone of modern agriculture. Gone are the days of rudimentary farming practices reliant solely on intuition and manual labor. Today, a sophisticated array of technological innovations stands poised to transform every facet of agricultural operations, from seed to harvest and beyond. At the heart of this technological revolution lies the paradigm of precision agriculture. Harnessing cutting-edge tools such as Global Positioning Systems (GPS), drones, sensors, and data analytics, precision agriculture empowers farmers to transcend the limitations of traditional farming methods. By meticulously mapping and monitoring every square inch of arable land, farmers can optimize resource allocation with surgical precision. From precisely calibrated irrigation systems that deliver water only where and when needed to targeted application of fertilizers and pesticides that minimize waste and environmental impact, precision agriculture epitomizes the marriage of efficiency and sustainability.

So, the data-driven decision-making lies at the core of modern farm management practices. In an era characterized by the deluge of information, farmers wield a powerful arsenal of data analytics tools to distill complex datasets into actionable insights. By leveraging real-time data on weather patterns, soil composition, crop health, and market dynamics, farmers can make informed decisions that optimize yields, mitigate risks and maximize profitability. Whether fine-tuning planting schedules, adjusting irrigation regimes or optimizing crop rotations, the ability to harness data empowers farmers to navigate the complexities of modern agriculture with unprecedented foresight and precision. Moreover, the advent of automation and robotics heralds a new era of efficiency and productivity in farm management. Automated machinery and robotic systems have revolutionized traditional farming operations, reducing labor costs and increasing operational efficiency. From autonomous tractors that plow fields with unparalleled accuracy to robotic harvesters that delicately pick fruits with surgical precision, the integration of automation technologies promises to redefine the boundaries of what is possible in agriculture.

Therefore, the role of technology in precise farm management represents more than mere technological innovation; it embodies a profound paradigm shift in our approach to agriculture. As we stand at the threshold of a new agricultural era, characterized by unprecedented challenges and opportunities, technology emerges as a potent catalyst for transformation. By embracing the principles of precision agriculture, data-driven decision-making and automation, farmers can transcend the limitations of conventional farming practices and forge a path towards a more sustainable, resilient and bountiful agricultural future.

A. Precision Agriculture

Precision agriculture, also known as precision farming or smart farming, is a modern approach to agriculture that utilizes technology and data-driven methods to optimize crop production while maximizing efficiency and sustainability. It represents a departure from traditional farming practices, which often rely on uniform application of inputs such as water, fertilizers, and pesticides across entire fields, regardless of variations in soil composition, crop health, or environmental conditions. Instead, precision agriculture seeks to precisely tailor inputs to the specific needs of individual plants or small areas within fields, thereby minimizing waste and environmental impact while maximizing yields and profitability.

At the core of precision agriculture is the concept of site-specific management, wherein farmers utilize a variety of technologies to gather detailed information about their fields and crops. These technologies include:

1. **Global Positioning Systems (GPS):** GPS technology allows farmers to precisely map the boundaries of their fields and track the location of farming equipment with pinpoint accuracy. This enables farmers to create detailed maps of soil types, topography, and other factors that influence crop growth and yield variability.
2. **Remote Sensing:** Remote sensing technologies, such as satellites, drones, and aerial imaging systems, provide farmers with detailed information about crop health, nutrient levels, water stress, and other important variables. By analyzing this data, farmers can identify areas of the field that may require additional inputs or interventions.
3. **Sensors and Monitoring Devices:** In-field sensors and monitoring devices measure soil moisture levels, temperature, humidity, and other environmental factors in real-time. This data allows farmers to make timely decisions about irrigation, fertilization, and pest management, optimizing resource use and minimizing waste.

4. **Variable Rate Technology (VRT):** VRT systems enable farmers to apply inputs such as fertilizers, pesticides, and irrigation water at variable rates across their fields, rather than using a uniform application approach. This allows farmers to match input levels to the specific needs of different areas within the field, resulting in more efficient resource utilization and improved crop performance.

5. **Data Analytics and Decision Support Systems:** Advanced data analytics software and decision support systems help farmers interpret the vast amounts of data collected from various sources. By analyzing historical data, current conditions, and predictive models, farmers can make informed decisions about crop management practices, input applications, and overall farm strategy.

The benefits of precision agriculture are manifold and extend to farmers, consumers and the environment alike. For farmers, precision agriculture offers increased efficiency, reduced input costs and improved profitability. By optimizing input use and minimizing waste, farmers can achieve higher yields and lower production costs, resulting in increased profits and a more sustainable farming operation. Consumers also benefit from precision agriculture through improved food quality, safety and consistency. By using precise input applications and monitoring techniques, farmers can produce higher-quality crops with fewer contaminants and pesticide residues, resulting in safer and more nutritious food products for consumers.

From an environmental perspective, precision agriculture offers significant advantages in terms of resource conservation and pollution reduction. By minimizing the use of fertilizers, pesticides, and water, precision agriculture helps to reduce nutrient runoff, soil erosion and groundwater contamination, leading to healthier ecosystems and improved water quality.

B. Data-driven Decision Making

Data-driven decision-making is a process in which businesses or organizations use data analysis and interpretation to guide their strategies, operations, and decision-making processes. In the context of agriculture, data-driven decision-making involves leveraging data from various sources, such as field sensors, satellite imagery, weather forecasts, market trends and historical records, to inform and optimize farming practices.

The process of data-driven decision-making in agriculture typically involves several key steps:

1. **Data Collection:** The first step is to gather relevant data from various sources. This may include collecting soil samples for analysis, installing sensors in the field to monitor soil moisture levels and weather conditions, using drones or satellites to capture aerial imagery of crop health and growth, and accessing market data to understand trends and demand.

2. **Data Processing and Analysis:** Once the data is collected, it needs to be processed and analyzed to extract meaningful insights. This may involve using statistical methods, machine learning algorithms, or other analytical techniques to identify patterns, correlations, and trends within the data.

3. **Decision Making:** Based on the insights gained from data analysis, farmers can make informed decisions about various aspects of their farming operation. This may include determining the optimal timing and amount of irrigation or fertilization, identifying areas of the field that require additional attention or intervention, adjusting planting or harvesting schedules based on weather forecasts or market conditions, and optimizing resource allocation to maximize productivity and profitability.

4. **Implementation and Monitoring:** After decisions are made, they need to be implemented in the field. This may involve adjusting equipment settings, applying inputs, or deploying specific management practices based on the decisions made. It's essential to monitor the outcomes of these decisions and collect additional data to evaluate their effectiveness continually.

5. **Iteration and Continuous Improvement:** Data-driven decision-making is an iterative process that involves continuously monitoring, analyzing, and refining farming practices based on new data and insights. By continually evaluating the outcomes of decisions and adjusting strategies accordingly, farmers can optimize their operations over time and adapt to changing conditions or requirements.

The benefits of data-driven decision-making in agriculture are significant. By leveraging data to inform their decisions, farmers can optimize resource use, increase crop yields, reduce input costs, minimize environmental impact, and improve overall farm profitability. Additionally, data-driven approaches enable farmers to respond more effectively to challenges such as climate variability, pest outbreaks, and market fluctuations, thereby increasing resilience and sustainability in agricultural production.

C. Automation and Robotics

Automation and robotics are revolutionizing precise farm management by introducing advanced technologies that streamline tasks, reduce labor costs and enhance efficiency across agricultural operations. In the context of farming, automation refers to the use of technology to perform tasks traditionally carried out by humans, while robotics involves the deployment of machines equipped with sensors, actuators, and artificial intelligence to perform specific agricultural tasks autonomously.

One of the primary applications of automation and robotics in precise farm management is in field operations such as planting, weeding and harvesting. Automated machinery, including self-driving tractors, seeders and harvesters, can navigate fields with precision, following predefined paths and using GPS technology to ensure accurate placement of seeds, fertilizers, and pesticides. This not only reduces the need for manual labor but also minimizes input waste and increases operational efficiency. The robotic systems equipped with computer vision and machine learning algorithms can identify and remove weeds with high accuracy, minimizing the use of herbicides and reducing crop losses due to weed competition. Additionally, robotic arms and grippers can be used for delicate tasks such as fruit picking, sorting, and packaging, improving productivity and reducing post-harvest losses.

Automation and robotics also play a crucial role in livestock management, where technologies such as automated feeding systems, robotic milkers and autonomous herding drones can optimize animal care and welfare while reducing labor requirements for farmers. These technologies can monitor animal health, track feeding and watering schedules, and even assist in disease detection and management, improving overall farm productivity and profitability.

Moreover, automation and robotics enable real-time monitoring and data collection, providing farmers with valuable insights into field conditions, crop health, and equipment performance. Sensors embedded in agricultural machinery can collect data on soil moisture levels, temperature, and nutrient concentrations, allowing farmers to make informed decisions about irrigation, fertilization, and pest control. This data-driven approach to farm management enables proactive problem-solving, reduces the risk of crop failure, and optimizes resource use.

D. Sustainable Agriculture

Sustainable agriculture is a holistic approach to farming that aims to meet the needs of the present without compromising the ability of future generations to meet their own needs. It encompasses a range of practices and principles designed to promote environmental stewardship, economic viability and social equity within the agricultural sector.

At its core, sustainable agriculture seeks to minimize the environmental impact of farming operations while maximizing the efficient use of resources such as water, soil and energy. This is achieved through the adoption of practices that enhance soil health, conserve water, reduce chemical inputs, and promote biodiversity. Some key components of sustainable agriculture include:

1. **Soil Conservation:** Sustainable agriculture prioritizes practices that maintain or improve soil health and fertility. This may include techniques such as conservation tillage, cover cropping, crop rotation, and the use of organic amendments. By preserving soil structure and organic matter, farmers can enhance soil resilience, reduce erosion, and improve long-term productivity.
2. **Water Management:** Sustainable agriculture emphasizes efficient water use and conservation strategies to minimize water waste and protect water quality. This may involve techniques such as drip irrigation, rainwater harvesting, soil moisture monitoring, and water-efficient crop selection. By optimizing irrigation practices and reducing runoff, farmers can mitigate water scarcity and minimize environmental impact.
3. **Biodiversity Conservation:** Sustainable agriculture recognizes the importance of biodiversity in supporting ecosystem health and resilience. Farmers may implement practices such as agroforestry, hedgerow planting, and the creation of wildlife habitats to enhance biodiversity on and around their farms. Maintaining diverse plant and animal populations helps to control pests and diseases, improve pollination, and promote ecological balance.
4. **Integrated Pest Management (IPM):** Sustainable agriculture promotes the use of IPM strategies to minimize reliance on synthetic pesticides and herbicides. IPM combines biological, cultural, and mechanical methods to control pests, weeds, and diseases while minimizing environmental impact and preserving beneficial organisms. By promoting natural pest control mechanisms and reducing chemical inputs, farmers can protect human health and ecosystem integrity.
5. **Agro-ecological Principles:** Sustainable agriculture is grounded in the principles of agro-ecology, which emphasize the interconnectedness of agricultural systems with ecological processes. Agro-ecological approaches seek to mimic natural ecosystems, leveraging ecological

principles to enhance farm productivity and resilience. By fostering synergies between crops, livestock, and the environment, farmers can optimize resource use and reduce reliance on external inputs.

In addition to environmental considerations, sustainable agriculture also addresses social and economic dimensions of farming, including labor rights, community development and economic viability for farmers. By promoting fair labor practices, equitable access to resources, and market opportunities for small-scale producers, sustainable agriculture contributes to the social and economic well-being of rural communities. Overall, sustainable agriculture represents a holistic and forward-thinking approach to farming that seeks to balance the needs of people, planet and profit. By embracing practices that promote environmental stewardship, economic viability, and social equity, sustainable agriculture offers a path towards a more resilient, equitable and sustainable food system for future generations.

E. Innovations in Crop Management

Innovations in crop management encompass a wide range of technologies and practices aimed at improving crop yields, reducing environmental impact, and enhancing sustainability in agriculture. These innovations leverage advancements in science, technology, and agronomy to address key challenges facing farmers, such as climate variability, soil degradation, pest and disease pressure and resource constraints. Here are some notable innovations in crop management:

1. **Genetic Engineering and Biotechnology:** Genetic engineering techniques such as genetic modification (GM) and gene editing enable scientists to develop crop varieties with desirable traits, such as resistance to pests and diseases, tolerance to environmental stresses like drought or salinity, and improved nutritional content. Biotechnology also includes innovations such as marker-assisted breeding, which accelerates the breeding process by identifying and selecting plants with desired traits at the molecular level.

2. **Precision Agriculture:** Precision agriculture utilizes technologies such as GPS, sensors, drones, and data analytics to optimize farm management practices with a high degree of precision. By collecting and analyzing data on soil conditions, crop health, weather patterns, and other variables, farmers can make informed decisions about planting, irrigation, fertilization, and pest management. This results in more efficient resource use, reduced input costs, and improved yields.

3. **Smart Irrigation Systems:** Smart irrigation systems utilize sensors, weather forecasts, and data analytics to deliver water to crops more efficiently and precisely. These systems can monitor soil moisture levels in real-time and adjust irrigation schedules accordingly, reducing water waste and minimizing the risk of overwatering or under watering. Smart irrigation technologies include drip irrigation, pivot irrigation, and soil moisture sensors.
4. **Biological Control and Integrated Pest Management (IPM):** Biological control methods harness natural enemies of pests, such as predatory insects, parasitic wasps, and beneficial microorganisms, to manage pest populations in crops. Integrated Pest Management (IPM) combines biological, cultural, and chemical control methods in a holistic approach to pest management, reducing reliance on synthetic pesticides and minimizing environmental impact while preserving natural pest control mechanisms.
5. **Crop Modeling and Predictive Analytics:** Crop modeling uses mathematical models and computer simulations to predict crop growth, development, and yield under different environmental conditions. By simulating the impact of factors such as temperature, precipitation, soil fertility, and management practices on crop performance, farmers can optimize planting decisions, input use, and crop rotations to maximize productivity and profitability.
6. **Vertical Farming and Controlled Environment Agriculture (CEA):** Vertical farming and CEA technologies enable the cultivation of crops in controlled indoor environments, such as greenhouses, hydroponic systems, and vertical stacks. These systems provide precise control over factors such as temperature, humidity, light, and nutrient levels, allowing for year-round production of high-quality crops with minimal environmental impact and resource use.

Therefore, the innovations in crop management hold great promise for addressing the challenges of feeding a growing global population while minimizing environmental impact and ensuring long-term sustainability in agriculture. By embracing advancements in genetics, technology, and agronomy, farmers can enhance crop yields, improve resource efficiency and contribute to a more resilient and sustainable food system for future generations.

F. Market Integration and Supply Chain Management

Market integration and supply chain management play critical roles in the agricultural sector, ensuring the efficient flow of goods from producers to consumers while maximizing value and minimizing waste. These processes involve coordinating the production, distribution, and marketing of agricultural products to meet consumer demand, optimize profitability, and promote sustainability. Here's an overview of market integration and supply chain management in agriculture:

1. **Market Analysis and Planning:** Market integration begins with analyzing consumer demand, market trends, and pricing dynamics to identify opportunities and risks. Farmers and agribusinesses use market research and forecasting tools to anticipate demand for specific crops, assess competition, and develop strategies for market entry and expansion. By understanding market dynamics, producers can make informed decisions about crop selection, pricing, and marketing strategies.
2. **Production Planning and Coordination:** Supply chain management starts at the farm level, where producers plan and coordinate production activities to meet market demand while optimizing resource use and minimizing risk. Farmers may use tools such as crop planning software, production scheduling, and contract farming agreements to coordinate planting schedules, input procurement, and labor management. By aligning production with market demand and quality standards, farmers can improve efficiency and profitability.
3. **Logistics and Distribution:** Once agricultural products are harvested, they must be transported efficiently from farms to processing facilities, distribution centers, and ultimately to retail outlets or consumers. Supply chain managers oversee logistics operations, including transportation, warehousing, and inventory management, to ensure timely delivery while minimizing transportation costs and spoilage. Advanced technologies such as GPS tracking, RFID tagging, and real-time monitoring systems help optimize logistics and streamline supply chain operations.
4. **Quality Assurance and Food Safety:** Ensuring the quality and safety of agricultural products is paramount in supply chain management. Farmers, processors, and distributors implement quality assurance programs, traceability systems, and food safety protocols to maintain product integrity and meet regulatory requirements. This includes monitoring product quality throughout the supply chain, implementing

hygiene practices, and conducting regular inspections and audits to identify and mitigate risks of contamination or spoilage.

5. **Market Access and Trade Facilitation:** Market integration also involves facilitating trade and market access for agricultural products, both domestically and internationally. Governments, industry associations, and trade organizations work to remove trade barriers, negotiate trade agreements, and harmonize regulatory standards to promote fair and open markets. This includes reducing tariffs, quotas, and non-tariff barriers to trade, as well as implementing measures to ensure product quality, safety, and compliance with international standards.
6. **Value Addition and Branding:** Supply chain management includes strategies to add value to agricultural products through processing, branding, and marketing initiatives. Farmers and agribusinesses may diversify their product offerings, develop branded products, or invest in value-added processing to differentiate their products and capture higher margins. This may involve developing new product lines, packaging innovations, or marketing campaigns to appeal to consumer preferences and create brand loyalty.

So, the market integration and supply chain management are essential components of the agricultural sector, enabling farmers and agribusinesses to efficiently produce, distribute and market agricultural products to meet consumer demand while maximizing value and sustainability. By optimizing supply chain operations, enhancing market access, and adding value through branding and innovation, stakeholders in the agricultural supply chain can contribute to a more resilient, efficient and sustainable food system.

Conclusion

The integration of technologies into precise farm management represents a paradigm shift in agriculture, offering solutions to the challenges of feeding a growing global population while mitigating environmental impact. These technologies empower farmers with real-time data, automation and precision tools, enabling them to optimize resource use, increase productivity and improve sustainability. From precision agriculture techniques like GPS-guided machinery and sensor-based monitoring to data-driven decision-making and automation, technological innovations are revolutionizing every aspect of farm management. By harnessing the power of data analytics, farmers can make informed decisions about inputs, irrigation and pest management, leading to higher yields and reduced environmental footprint. Furthermore, automation and robotics streamline labor-intensive tasks, reducing reliance on manual labor and increasing operational efficiency. These advancements not only improve farm productivity but also free up farmers to focus on strategic decision-making and innovation. As we navigate the complexities of modern agriculture, the role of technology in precise farm management will continue to evolve, driving further advancements in productivity, sustainability, and resilience. By embracing these technologies, farmers can unlock new opportunities for growth, ensuring a more secure and sustainable food supply for future generations.

References

Ardito, L., Petruzzelli, A.M., Panniello, U. and Garavelli, A.C. 2018. Towards Industry 4.0: Mapping digital technologies for supply chain management-marketing integration. Business Process Management Journal, 25(2): 323-346.

Balafoutis, A., Beck, B., Fountas, S., Vangeyte, J., Van der Wal, T., Soto, I., Gómez-Barbero, M., Barnes, A. and Eory, V. 2017. Precision agriculture technologies positively contributing to GHG emissions mitigation, farm productivity and economics. Sustainability, 9(8): 1339.

Beriya, A. and Saroja, V.N. 2019. Data-Driven Decision Making for Smart Agriculture (No. 8). ICT India Working Paper.

Billingsley, J., Visala, A. and Dunn, M. 2008. Robotics in agriculture and forestry.

Hedley, C. 2015. The role of precision agriculture for improved nutrient management on farms. Journal of the Science of Food and Agriculture, 95(1): 12-19.

Kulkarni, A.A., Dhanush, P., Chetan, B.S., Gowda, C.T. and Shrivastava, P.K. 2020. Applications of automation and robotics in agriculture industries; a review. In IOP Conference Series: Materials Science and Engineering, 748(1): 012002. IOP Publishing.

Mahmud, M.S.A., Abidin, M.S.Z., Emmanuel, A.A. and Hasan, H.S. 2020. Robotics and automation in agriculture: present and future applications. Applications of Modelling and Simulation, 4: 130-140.

Njoroge, B.M., Fei, T.K. and Thiruchelvam, V. 2018. A research review of precision farming techniques and technology. J. Appl. Technol. Innov, 2(9).

Power, D. 2005. Supply Chain Management Integration and Implementation: a literature review. Supply Chain Management: an International Journal, 10(4): 252-263.

Rehman, A., Jingdong, L., Khatoon, R., Hussain, I. and Iqbal, M.S. 2016. Modern agricultural technology adoption its importance, role and usage for the improvement of agriculture. Life Science Journal, 14(2): 70-74.

Rozenstein, O., Cohen, Y., Alchanatis, V., Behrendt, K., Bonfil, D.J., Eshel, G., Harari, A., Harris, W.E., Klapp, I., Laor, Y. and Linker, R. 2024. Data-driven agriculture and sustainable farming: friends or foes?. Precision Agriculture, 25(1): 520-531.

Ruben, R., van Tilburg, A., Trienekens, J. and van Boekel, M. 2007. Linking market integration, supply chain governance, quality and value added in tropical food chains. Tropical food chains: governance regimes for quality management, 13-46.

Tantalaki, N., Souravlas, S. and Roumeliotis, M. 2019. Data-driven decision making in precision agriculture: The rise of big data in agricultural systems. Journal of Agricultural & Food Information, 20(4): 344-380.

2

The Influence of Government Policies on Agricultural Practices

***Ankit Kumar Maurya*[1]*, Ritesh Singh Gangwar*[2]**
***Rahul Kumar Rai*[3]**

[1]*Agricultural Economics, Chandra Shekhar Azad University of Agriculture and Technology, Kanpur, U.P.*
[2]*Subject Matter Specialist (Agronomy), KVK, Chandauli, U.P.*
[3]*Agricultural Economics, Banda University of Agriculture and Technology Banda, U.P.*

Abstract

Government policies wield a profound influence on agricultural practices, shaping the landscape of global food production, environmental sustainability, and economic viability. This paper examines the multifaceted impact of governmental interventions on agricultural practices across diverse domains. Subsidies and support programs play a pivotal role, incentivizing farmers to adopt specific practices such as crop diversification or conservation methods. Trade policies, including tariffs and trade agreements, dictate market access and competitiveness, steering agricultural practices towards certain production methods. Regulatory frameworks ensure food safety, environmental protection, and land use sustainability, guiding farmers' decisions and behaviours. Research and development funding propel innovation in agriculture, driving the adoption of new technologies and sustainable practices. Land use policies and environmental regulations influence land management decisions, impacting agricultural practices with implications for conservation and resource management. Moreover, risk management programs and labour policies mitigate uncertainties and labour challenges, shaping farm operations and productivity. Energy policies and social welfare programs further intersect with agricultural practices, influencing input costs, energy usage, and consumer demand. Understanding the intricate interplay between government policies and agricultural practices is essential for fostering sustainable food systems, enhancing resilience, and addressing global challenges such as climate change and food security. This paper underscores the importance of coherent and adaptive policy frameworks that balance economic, social, and environmental imperatives to foster a resilient and sustainable agricultural sector.

Keywords: *Policies, Food Production, Market Access, Resource Management, etc.*

Introduction

The influence of government policies on agricultural practices is a topic of paramount importance in the realm of global food security, environmental sustainability and economic development. Governments around the world play a pivotal role in shaping the trajectory of agricultural systems through a myriad of interventions, regulations, and support mechanisms. The intricate interplay between governmental policies and agricultural practices underscores the complex dynamics that govern food production, land use, and rural livelihoods. Agriculture stands as a cornerstone of human civilization, providing sustenance, employment, and economic prosperity to billions of people worldwide. However, the contemporary agricultural landscape is characterized by a myriad of challenges ranging from climate change and resource depletion to food insecurity and rural poverty. In this context, government policies emerge as crucial instruments for steering agricultural practices towards sustainability, resilience, and equitable development.

Subsidies and support programs constitute one of the primary mechanisms through which governments influence agricultural practices. By providing financial incentives, subsidies, and technical assistance, governments seek to promote specific practices such as crop diversification, soil conservation, and adoption of sustainable farming methods. These policies not only influence farmers' decision-making but also shape the overall structure of agricultural markets and supply chains. Trade policies represent another key dimension of governmental influence on agricultural practices. Tariffs, quotas, and trade agreements dictate market access, price dynamics, and competitiveness, thereby influencing farmers' production choices and market orientation. Moreover, trade policies have profound implications for the adoption of certain production methods, as trade liberalization may incentivize intensification or specialization in certain commodities, potentially compromising environmental sustainability and food security. Regulatory frameworks constitute a critical tool for governments to ensure food safety, environmental protection, and land use sustainability within the agricultural sector. Regulations governing pesticide use, water management, and land conservation influence farmers' practices and behaviours, shaping the environmental footprint of agriculture and mitigating risks to public health and ecosystem integrity.

Furthermore, government investments in research and development (R&D) play a pivotal role in driving innovation and technological advancement in agriculture. By funding agricultural research, extension services, and technology transfer initiatives, governments facilitate the adoption of new technologies, crop varieties, and management practices aimed at enhancing productivity, resilience, and sustainability within the agricultural sector. Land use policies represent another key dimension of governmental influence on agricultural practices. Governments regulate land ownership, land tenure systems, and zoning regulations, thereby shaping land use patterns, access to resources, and incentives for conservation or conversion of agricultural land. These policies have profound implications for rural development, biodiversity conservation, and resilience to environmental change. Moreover, governmental interventions in risk management and labour policies significantly impact agricultural practices and productivity. Risk management programs such as crop insurance and disaster assistance help farmers mitigate the financial risks associated with agricultural production, influencing decision-making regarding crop selection, input use, and production methods. Labour policies, including regulations related to wages, working conditions, and immigration, shape labour dynamics within the agricultural sector, influencing production costs, labour availability, and farm management practices.

Overall, the influence of government policies on agricultural practices encompasses a diverse array of dimensions, ranging from subsidies and trade policies to regulations, research funding, and land use planning. Understanding the multifaceted nature of governmental interventions in agriculture is crucial for fostering sustainable food systems, enhancing resilience, and addressing global challenges such as climate change and food insecurity. This paper aims to explore and analyse the intricate interplay between government policies and agricultural practices, highlighting the opportunities and challenges for promoting a resilient, equitable, and sustainable agricultural sector in the 21st century.

Subsidies and Support Programs

Subsidies and support programs constitute integral components of government policies that profoundly influence agricultural practices, production decisions, and market dynamics. These programs encompass a diverse array of initiatives aimed at incentivizing specific practices, supporting farmers' incomes, and achieving broader policy objectives such as food security, rural development, and environmental sustainability.

1. **Income Stabilization and Risk Mitigation:** One primary objective of subsidies and support programs is to stabilize farmers' incomes and mitigate financial risks associated with agricultural production. Direct payments, income support schemes, and crop insurance programs provide financial assistance to farmers, particularly during periods of market volatility, price fluctuations, or production losses. By ensuring income stability, these programs enable farmers to make long-term investments, adopt sustainable practices, and withstand economic uncertainties, thereby promoting stability and resilience within the agricultural sector.
2. **Incentivizing Desired Practices:** Subsidies and support programs often target specific agricultural practices deemed desirable for achieving broader policy goals such as environmental conservation, resource management, and market competitiveness. Governments provide financial incentives, technical assistance, and research funding to encourage farmers to adopt practices such as conservation tillage, organic farming, and agroforestry. By rewarding sustainable practices, subsidies and support programs promote soil health, water conservation, biodiversity conservation, and climate resilience, contributing to the long-term sustainability of agricultural systems.
3. **Market and Trade Support:** In addition to income stabilization and practice incentivization, subsidies and support programs may also include measures to support farmers' access to markets, facilitate trade, and enhance competitiveness. Price support mechanisms, export subsidies, and marketing assistance programs aim to stabilize prices, ensure market access, and promote agricultural exports, thereby bolstering farmers' incomes and market opportunities. However, these programs can also distort market incentives, create trade tensions, and impact global commodity prices, necessitating careful policy design and international cooperation.
4. **Social and Rural Development:** Subsidies and support programs often play a crucial role in promoting social welfare, rural development, and poverty alleviation in agricultural communities. Governments may provide subsidies for input costs such as seeds, fertilizers, and machinery, or invest in rural infrastructure, education, and healthcare services to enhance the quality of life for rural populations. By supporting rural livelihoods and fostering economic development, these programs contribute to social cohesion, poverty reduction, and equitable growth within agricultural regions.

Overall, subsidies and support programs represent powerful tools for governments to shape agricultural practices, achieve policy objectives, and support the livelihoods of farmers. While these programs play a vital role in stabilizing incomes, incentivizing sustainable practices, and promoting rural development, they also raise challenges such as market distortions, budget constraints, and environmental impacts. Balancing the imperatives of support programs with the need for efficiency, equity, and sustainability remains a key challenge for policymakers seeking to foster resilient and inclusive agricultural systems.

Trade Policies

Trade policies constitute a significant aspect of government intervention in agriculture, shaping production decisions, market access, and competitiveness within the agricultural sector. Governments employ a range of trade policies, including tariffs, quotas, subsidies, and trade agreements, to regulate the flow of agricultural goods across borders and safeguard domestic interests. These policies have profound implications for agricultural practices, market dynamics, and the overall structure of the agricultural sector.

1. **Market Access and Competitiveness:** One primary objective of trade policies is to facilitate market access for agricultural producers while safeguarding domestic markets from foreign competition. Tariffs, quotas, and non-tariff barriers are used to regulate imports, protect domestic producers, and maintain market stability. By controlling market access, governments influence production decisions, investment patterns, and the competitiveness of domestic agriculture. Moreover, trade policies impact the distribution of benefits and costs across different sectors of the economy, affecting farmers' incomes, consumer prices, and economic welfare.
2. **Export Promotion and Subsidies:** Trade policies also include measures aimed at promoting agricultural exports, enhancing competitiveness, and capturing a larger share of global markets. Governments may provide export subsidies, trade finance, and marketing assistance to support agricultural exporters, reduce trade barriers, and stimulate export growth. However, export-oriented subsidies may distort international trade, create trade tensions, and impact global commodity prices, necessitating careful policy coordination and international cooperation.
3. **Regulatory Harmonization and Standards:** Trade agreements often include provisions related to sanitary and phytosanitary standards, food safety regulations, and intellectual property rights, which can influence

agricultural practices and market access. Governments harmonize regulations, adopt common standards, and establish mutual recognition agreements to facilitate trade, enhance food safety, and promote consumer confidence in agricultural products. Moreover, trade agreements may impact the adoption of sustainable agricultural practices by influencing market demand for certified products, organic goods, or products with specific environmental attributes.

4. **Market Volatility and Risk Management:** Trade policies can exacerbate market volatility and uncertainty for agricultural producers by influencing supply and demand dynamics, price fluctuations, and trade patterns. Governments implement risk management measures such as import tariffs, export restrictions, and strategic reserves to stabilize markets, mitigate price volatility, and protect domestic producers from external shocks. However, trade-related risks and market uncertainties pose challenges for farmers, agribusinesses, and policymakers seeking to navigate the complexities of global trade and agricultural production.

So, trade policies exert significant influence over agricultural practices, market dynamics, and the overall structure of the agricultural sector. While these policies can promote market access, competitiveness, and export growth, they also raise challenges such as market volatility, trade tensions, and regulatory complexities. Balancing the imperatives of trade policy with the need for sustainable and equitable agricultural development remains a key challenge for policymakers seeking to foster resilient and inclusive agricultural systems in an interconnected global economy.

Regulatory Frameworks

Regulatory frameworks established by governments play a critical role in influencing agricultural practices by setting standards, requirements, and guidelines for farmers, agribusinesses, and other stakeholders within the agricultural sector. These regulations aim to ensure food safety, environmental protection, animal welfare, and public health while also addressing broader societal concerns related to land use, biodiversity conservation, and sustainable development. One of the primary areas where regulatory frameworks impact agricultural practices is in the realm of food safety and quality standards. Governments establish regulations governing the use of pesticides, fertilizers, and other agrochemicals to protect consumers from harmful residues and ensure the safety of food products. Similarly, regulations related to food labelling, packaging, and processing standards aim to provide consumers with accurate information and protect against fraud or misrepresentation in the marketplace.

Environmental regulations constitute another key dimension of regulatory frameworks that influence agricultural practices. Governments enact laws and policies to mitigate pollution, conserve natural resources, and promote sustainable land management practices within the agricultural sector. Regulations governing water use, soil conservation, and habitat protection seek to minimize the environmental impacts of agricultural activities and preserve ecosystem integrity for future generations. Moreover, regulatory frameworks address animal welfare concerns by establishing standards for the treatment and handling of livestock in agricultural operations. Regulations governing livestock housing, transportation, and slaughter practices aim to ensure humane treatment and minimize stress and suffering for animals raised for food production. In addition to food safety, environmental protection, and animal welfare, regulatory frameworks also encompass land use planning, zoning regulations, and land tenure systems that influence agricultural practices. Governments enact policies to regulate land conversion, protect agricultural land from urban sprawl, and promote sustainable land use practices such as agroforestry, conservation agriculture, and integrated landscape management.

Overall, regulatory frameworks represent a crucial tool for governments to influence agricultural practices and achieve various policy objectives related to food security, environmental sustainability, and public health. By setting standards, requirements, and guidelines for agricultural activities, governments seek to balance the interests of farmers, consumers, and society at large while promoting a resilient, equitable, and sustainable agricultural sector. However, achieving this balance requires continuous monitoring, enforcement, and adaptation of regulatory frameworks to address emerging challenges and opportunities within the dynamic agricultural landscape.

Research and Development Funding

Research and development (R&D) funding represents a pivotal aspect of governmental influence on agricultural practices, driving innovation, technological advancement, and productivity enhancement within the sector. Governments allocate resources to support agricultural R&D initiatives, including basic research, applied research, technology transfer, and extension services, aiming to address key challenges such as climate change, food security, and sustainability.

1. **Fostering Innovation:** Government-funded R&D programs serve as catalysts for innovation within the agricultural sector by supporting scientific research, experimentation, and technology development. Research institutions, universities, and public-private partnerships receive

funding to conduct studies on various aspects of agriculture, including crop genetics, soil health, pest management, and water conservation. These efforts lead to the development of new technologies, tools, and practices that enhance the efficiency, resilience, and sustainability of agricultural production systems.

2. **Technology Transfer and Extension Services:** In addition to funding research activities, governments invest in technology transfer and extension services to disseminate research findings and best practices to farmers, agribusinesses, and rural communities. Extension agents, agricultural advisors, and outreach programs provide education, training, and technical assistance to help farmers adopt innovative technologies and management practices. By bridging the gap between research and practice, extension services facilitate the adoption of sustainable agricultural practices, increasing productivity and profitability while minimizing environmental impacts.

3. **Addressing Emerging Challenges:** Government-funded R&D initiatives play a crucial role in addressing emerging challenges facing the agricultural sector, such as climate change, resource scarcity, and pest outbreaks. Researchers receive funding to develop climate-resilient crop varieties, drought-tolerant farming systems, and precision agriculture technologies that optimize resource use and mitigate environmental risks. Moreover, R&D funding supports studies on integrated pest management, biosecurity measures, and disease surveillance, helping farmer's combat pests and diseases while reducing reliance on chemical inputs.

4. **Promoting Collaboration and Partnerships:** Governmental R&D funding encourages collaboration and partnerships among researchers, industry stakeholders, and international organizations, fostering knowledge exchange, technology transfer, and capacity building. Research consortia, public-private partnerships, and international networks receive funding to tackle complex agricultural challenges through interdisciplinary approaches and shared resources. By promoting collaboration, R&D funding enhances the effectiveness and impact of agricultural research efforts, leading to breakthrough innovations and sustainable solutions.

Therefore, government-funded R&D initiatives play a pivotal role in catalysing innovation, technology adoption, and sustainability within the agricultural sector. By investing in research, technology transfer, and collaboration, governments contribute to the development of resilient, productive, and environmentally sustainable agricultural systems that meet the evolving needs of society while addressing global challenges such as food security and climate change.

Land Use Policies

Land use policies implemented by governments play a crucial role in influencing agricultural practices, land management decisions, and environmental sustainability within the agricultural sector. These policies encompass a range of measures aimed at regulating land use, preserving natural resources, and promoting sustainable development in rural areas.

1. **Zoning Regulations and Land Tenure Systems:** One key aspect of land use policies is zoning regulations and land tenure systems, which govern land ownership, land use planning, and property rights within agricultural regions. Governments establish zoning regulations to designate land for agricultural, residential, industrial, or conservation purposes, shaping land use patterns and development activities. Moreover, land tenure systems such as land redistribution, land leasing, and land titling influence farmers' access to land, investment incentives, and agricultural productivity.

2. **Conservation and Environmental Protection:** Land use policies also include measures to conserve natural resources, protect biodiversity, and preserve ecosystems within agricultural landscapes. Governments enact regulations governing soil conservation, water management and habitat protection to minimize environmental degradation, enhance ecosystem services, and promote resilience to climate change. Conservation programs, Agri-environment schemes, and land stewardship initiatives provide incentives for farmers to adopt sustainable land management practices such as agroforestry, conservation tillage, and riparian buffers.

3. **Agricultural Land Preservation:** In addition to environmental conservation, land use policies aim to preserve agricultural land from urbanization, fragmentation, and conversion to non-agricultural uses. Governments implement policies such as agricultural land trusts, agricultural zoning ordinances, and development restrictions to protect farmland from urban sprawl and ensure its availability for agricultural production. These policies support the viability of farming operations, maintain rural landscapes, and safeguard food security by preserving productive agricultural land for future generations.

4. **Land Use Planning and Infrastructure Development:** Land use policies also encompass land use planning, infrastructure development, and rural investment initiatives aimed at promoting sustainable land use practices and rural development. Governments invest in rural infrastructure such as roads, irrigation systems, and market facilities to enhance access to markets, reduce transportation costs, and improve the quality of life for rural communities. Moreover, land use planning efforts integrate agricultural development with broader economic, social, and environmental objectives, fostering integrated landscape management and sustainable land use practices.

Overall, land use policies exert significant influence over agricultural practices, land management decisions, and environmental sustainability within the agricultural sector. By regulating land use, preserving natural resources, and promoting sustainable development, governments contribute to the resilience, productivity, and long-term viability of agricultural systems. However, achieving sustainable land use outcomes requires coordination, collaboration, and adaptive governance approaches that balance competing interests and address emerging challenges in an increasingly complex and interconnected world.

Environmental Policies

Environmental policies implemented by governments play a crucial role in shaping agricultural practices, promoting environmental sustainability, and mitigating the environmental impacts of agricultural production. These policies encompass a range of measures aimed at conserving natural resources, reducing pollution, and fostering resilience to environmental challenges within the agricultural sector.

1. **Regulation of Pesticide and Fertilizer Use:** One key aspect of environmental policies is the regulation of pesticide and fertilizer use to minimize environmental contamination and protect human health. Governments establish regulations governing the registration, sale, and application of pesticides and fertilizers, including restrictions on toxic chemicals, application rates, and buffer zones near water bodies. By promoting integrated pest management (IPM) practices, organic farming methods, and precision agriculture technologies, environmental policies aim to reduce reliance on chemical inputs and mitigate risks to soil, water, and biodiversity.

2. **Water Management and Conservation:** Environmental policies also address water management and conservation within agricultural

landscapes, aiming to optimize water use efficiency, protect water quality, and mitigate water-related risks such as droughts and floods. Governments enact regulations governing water withdrawals, irrigation practices, and water pollution control to safeguard freshwater resources and promote sustainable water management practices in agriculture. Moreover, water conservation programs, irrigation efficiency incentives, and watershed management initiatives encourage farmers to adopt water-saving technologies and practices, enhancing resilience to water scarcity and climate variability.

3. **Soil Conservation and Land Stewardship:** In addition to water management, environmental policies focus on soil conservation and land stewardship to minimize soil erosion, enhance soil health, and preserve ecosystem services within agricultural ecosystems. Governments implement regulations governing soil conservation practices, erosion control measures, and land use planning to promote sustainable land management practices such as conservation tillage, cover cropping, and agroforestry. Furthermore, soil conservation programs, land stewardship incentives, and Agri-environment schemes provide financial support and technical assistance to farmers to adopt practices that improve soil quality, enhance carbon sequestration, and protect natural habitats.

4. **Biodiversity Conservation and Habitat Protection:** Environmental policies also prioritize biodiversity conservation and habitat protection within agricultural landscapes, aiming to preserve ecosystem diversity, support pollinators, and maintain wildlife habitats. Governments establish regulations governing habitat preservation, endangered species protection, and biodiversity offsets to mitigate the adverse impacts of agricultural activities on biodiversity and ecosystem integrity. Moreover, biodiversity conservation programs, wildlife habitat restoration initiatives, and agro ecological approaches promote the integration of biodiversity conservation with agricultural production, fostering ecological resilience and enhancing the sustainability of agricultural systems.

Therefore, environmental policies play a crucial role in promoting sustainable agricultural practices, reducing environmental impacts, and fostering resilience to environmental challenges within the agricultural sector. By regulating pesticide and fertilizer use, promoting water management and conservation, encouraging soil conservation and land stewardship, and prioritizing biodiversity conservation and habitat protection, governments contribute to the long-term viability and environmental sustainability of agricultural systems.

However, achieving sustainable agricultural outcomes requires collaborative efforts, adaptive management approaches, and stakeholder engagement to address complex environmental issues and ensure the coexistence of agriculture and the environment.

Risk Management Programs

Risk management programs implemented by governments play a crucial role in mitigating the financial uncertainties and challenges faced by farmers within the agricultural sector. These programs encompass a range of initiatives aimed at protecting farmers' incomes, enhancing resilience to production risks, and promoting stability within agricultural markets.

1. **Crop Insurance and Disaster Assistance:** One primary component of risk management programs is crop insurance and disaster assistance, which provide financial protection to farmers against crop losses due to natural disasters, adverse weather events, or other unforeseen circumstances. Governments offer subsidized crop insurance programs that indemnify farmers for yield losses, revenue shortfalls, or input cost increases resulting from weather-related disasters, pests, diseases, or market fluctuations. Moreover, disaster assistance programs provide emergency relief and financial support to farmers affected by severe weather events, such as floods, droughts, hurricanes, or wildfires, helping them recover from production losses and mitigate economic hardships.
2. **Price Stabilization and Income Support:** Risk management programs also include measures to stabilize prices, manage market volatility, and support farmers' incomes during periods of low commodity prices or oversupply. Governments implement price support mechanisms, market intervention programs, and income stabilization initiatives to buffer farmers against price fluctuations, ensure a minimum level of income, and maintain market stability. These programs may involve government purchases of surplus commodities, price floors, or income support payments to farmers facing financial distress or market uncertainty.
3. **Input Subsidies and Financial Assistance:** In addition to crop insurance and income support, risk management programs may include input subsidies, credit facilities, and financial assistance to help farmers manage input costs, access capital, and invest in productivity-enhancing technologies or practices. Governments provide subsidies for agricultural inputs such as seeds, fertilizers, pesticides, and machinery, reducing production costs and enhancing farmers' competitiveness. Moreover, credit programs, loan guarantees, and microfinance initiatives facilitate

access to affordable credit and financial services, enabling farmers to make investments, manage risks, and improve their livelihoods.

4. **Market Diversification and Risk Mitigation:** Risk management programs also promote market diversification, risk mitigation strategies, and alternative income sources to reduce farmers' reliance on a single commodity or market outlet. Governments support diversification efforts through extension services, technical assistance, and financial incentives for crop rotation, value-added production, or alternative income-generating activities such as agro-tourism, agroforestry, or direct marketing. By encouraging diversification and innovation, risk management programs enhance farmers' resilience to market volatility, production risks, and economic uncertainties, fostering long-term sustainability within the agricultural sector.

In conclusion, risk management programs constitute essential tools for governments to safeguard agricultural livelihoods, enhance resilience, and promote stability within the agricultural sector. By providing crop insurance, disaster assistance, price stabilization measures, and financial support, governments help farmers manage production risks, overcome financial challenges, and withstand external shocks. However, effective risk management requires comprehensive approaches, stakeholder engagement, and adaptive strategies to address evolving threats and ensure the viability of agricultural livelihoods in an uncertain and dynamic environment.

Labour Policies

Labour policies enacted by governments play a significant role in influencing agricultural practices, workforce dynamics, and rural livelihoods within the agricultural sector. These policies encompass a range of measures aimed at regulating labour relations, ensuring fair wages and working conditions, and addressing labour shortages and migration within agricultural communities.

1. **Regulation of Wages and Working Conditions:** One key aspect of labour policies is the regulation of wages, working hours, and employment conditions to protect the rights and well-being of agricultural workers. Governments establish minimum wage laws, overtime regulations, and workplace safety standards to ensure fair compensation and safe working conditions for farm labourers. Moreover, labour laws may include provisions for health insurance, social security benefits, and unemployment insurance to provide financial protection and social support to agricultural workers.

2. **Seasonal and Migrant Labour Programs:** Labour policies also address seasonal labour needs and migrant worker programs within the agricultural sector, particularly in regions with seasonal production cycles or labour shortages. Governments implement seasonal labour programs, guest worker programs, or temporary foreign worker initiatives to facilitate the recruitment and employment of migrant workers for seasonal agricultural tasks such as planting, harvesting, and processing. These programs provide farmers with access to a reliable labour force while offering employment opportunities and income support to migrant workers.

3. **Labour Rights and Social Protections:** In addition to regulating wages and working conditions, labour policies promote labour rights, collective bargaining, and social protections for agricultural workers. Governments establish labour standards, labour unions, and labour inspection systems to enforce compliance with labour laws and protect workers from exploitation, discrimination, and abuse. Moreover, social protection programs such as unemployment benefits, healthcare services, and housing assistance help agricultural workers access essential services and mitigate economic hardships.

4. **Training and Skill Development:** Labour policies also focus on training, skill development, and workforce education to enhance the productivity, efficiency, and professionalism of agricultural labourers. Governments invest in vocational training programs, agricultural extension services, and skills development initiatives to equip workers with the knowledge, expertise, and technical skills required to perform agricultural tasks effectively and safely. Moreover, apprenticeship programs, on-the-job training, and career advancement opportunities provide pathways for agricultural workers to improve their skills, expand their employment options, and pursue long-term career goals within the agricultural sector.

Overall, labour policies play a crucial role in shaping agricultural practices, workforce dynamics, and rural livelihoods within the agricultural sector. By regulating wages and working conditions, addressing seasonal labour needs, protecting labour rights, and promoting workforce training and skill development, governments contribute to the well-being, stability, and sustainability of agricultural communities and the agricultural workforce. However, ensuring effective implementation, enforcement, and compliance with labour laws remains essential to safeguarding the rights and dignity of agricultural workers and fostering inclusive and equitable development within the agricultural sector.

Energy Policies

Energy policies implemented by governments play a crucial role in shaping agricultural practices, resource use, and sustainability within the agricultural sector. These policies encompass a range of measures aimed at promoting energy efficiency, renewable energy adoption, and sustainable resource management to enhance productivity, reduce environmental impacts, and mitigate climate change.

1. **Promoting Renewable Energy Adoption:** One key aspect of energy policies is the promotion of renewable energy adoption within the agricultural sector, including solar, wind, biomass, and bioenergy technologies. Governments offer incentives, subsidies, and tax credits to encourage farmers to invest in renewable energy systems such as solar panels, wind turbines, and biogas digesters to generate clean energy and reduce reliance on fossil fuels. Moreover, renewable energy programs, feed-in tariffs, and net metering policies facilitate grid integration and promote distributed energy generation, enabling farmers to generate electricity onsite and contribute to renewable energy production.

2. **Enhancing Energy Efficiency:** Energy policies also focus on enhancing energy efficiency and reducing energy consumption within agricultural operations through technological innovation, best practices, and efficiency standards. Governments provide grants, loans, and technical assistance to support energy efficiency upgrades, equipment retrofits, and adoption of energy-saving technologies such as energy-efficient lighting, irrigation systems, and machinery. Moreover, energy efficiency programs, energy audits, and energy management training help farmers identify opportunities for energy savings, optimize resource use, and improve operational efficiency.

3. **Biofuel Production and Biomass Utilization:** In addition to renewable energy, energy policies promote biofuel production and biomass utilization as sustainable alternatives to fossil fuels within the agricultural sector. Governments incentivize biofuel production from agricultural feed stocks such as corn, sugarcane, and oilseeds through biofuel mandates, tax incentives, and subsidies. Moreover, biomass utilization programs, crop residue management initiatives, and biomass conversion technologies promote the sustainable use of agricultural residues, crop residues, and organic waste for energy production, heat generation, and bio-based products, reducing greenhouse gas emissions and promoting circular economy principles.

4. **Research and Development:** Energy policies also prioritize research and development (R&D) funding to drive innovation, technology adoption, and sustainability within the agricultural sector. Governments invest in R&D programs, demonstration projects, and technology transfer initiatives to develop advanced energy technologies, renewable energy solutions, and sustainable agricultural practices. Moreover, research partnerships, academic collaborations, and knowledge exchange platforms facilitate the dissemination of best practices, technical expertise, and innovative solutions to farmers, agribusinesses, and rural communities.

Social Welfare Programs

Government policies wield significant influence over agricultural practices, often shaping the socioeconomic landscape of rural communities. In this context, social welfare programs play a pivotal role in supporting farmers and addressing issues of food security, poverty, and rural development. Here's a concise overview of how government policies intersect with agricultural practices and social welfare programs:

1. **Subsidies and Incentives:** Governments frequently offer subsidies and incentives to farmers to encourage specific agricultural practices. These may include subsidies for fertilizers, seeds, machinery, or irrigation systems. Social welfare programs often complement these initiatives by providing financial assistance or grants to small-scale farmers, enabling them to adopt modern technologies and improve productivity.

2. **Price Support Mechanisms:** Price support mechanisms such as minimum support prices (MSPs) ensure farmers receive a fair price for their produce. Government procurement at MSPs stabilizes agricultural markets and protects farmers from price fluctuations. Social welfare programs may intervene by providing direct cash transfers or food subsidies to vulnerable farming communities during periods of price volatility or crop failure.

3. **Agricultural Extension Services:** Government-funded agricultural extension services disseminate knowledge, technologies, and best practices among farmers. These services promote sustainable farming methods, efficient resource utilization, and crop diversification. Social welfare programs often collaborate with extension services to deliver targeted training and capacity-building initiatives, empowering farmers to improve their livelihoods and resilience to external shocks.

4. **Land Reform and Tenure Security:** Government policies on land reform and tenure security impact agricultural practices and rural livelihoods. Secure land tenure encourages long-term investments in soil conservation, agroforestry, and sustainable land management. Social welfare programs may support landless farmers through land redistribution initiatives, land titling programs, or land leasing arrangements, enhancing their access to productive resources and social protection.
5. **Environmental Conservation and Climate Resilience:** Government regulations and incentives aimed at promoting environmental conservation and climate resilience influence agricultural practices. Agro ecological approaches, conservation agriculture, and climate-smart farming techniques are encouraged through policy frameworks. Social welfare programs often integrate environmental conservation objectives, providing incentives for eco-friendly farming practices and supporting adaptation measures to mitigate climate risks faced by farmers.

Conclusion

In essence, the nexus between government policies and agricultural practices underscores a crucial interplay that shapes the livelihoods of millions worldwide. Through subsidies, price support mechanisms, extension services, land reforms, and environmental initiatives, governments wield significant influence over farming communities, steering them towards sustainable practices and bolstering their resilience. Yet, the efficacy of these policies hinges on their alignment with broader development goals and the inclusivity of social welfare programs. Only when policies prioritize the needs of small-scale farmers, address systemic inequalities, and foster community participation can they truly catalyse transformative change. Ultimately, the impact of government policies extends far beyond agricultural productivity; it reverberates through entire rural economies, impacting food security, poverty alleviation, and environmental sustainability. Thus, policymakers must adopt a holistic approach, considering the multifaceted challenges faced by farmers and the intricate dynamics of rural communities. In conclusion, the influence of government policies on agricultural practices is profound and far-reaching, shaping not only the way we produce food but also the social fabric of rural societies. By embracing a comprehensive and inclusive approach, governments can harness agriculture's potential as a driver of sustainable development, ensuring the prosperity and well-being of present and future generations.

References

Azadi, H., Petrescu, D.C., Petrescu-Mag, R.M. and Ozunu, A. 2020. Environmental risk mitigation for sustainable land use development. Land Use Policy, 95: 104488.

Conrad, K. 1993. Taxes and subsidies for pollution-intensive industries as trade policy. Journal of environmental economics and management, 25(2): 121-135.

Goldstein, A.P. and Kearney, M. 2020. Know when to fold 'em: An empirical description of risk management in public research funding. Research Policy, 49(1): 103873.

Greening, L.A. and Bernow, S. 2004. Design of coordinated energy and environmental policies: use of multi-criteria decision-making. Energy policy, 32(6): 721-735.

Hall, A., Mytelka, L. and Oyeyinka, B. 2005. Innovation systems: Implications for agricultural policy and practice.

Hobbs, J.E. 2003. Incentives for the Adoption of Good Agricultural Practices (GAPs). Food and Agriculture Organization, 1.

Holzmann, R. and Jørgensen, S. 2001. Social risk management: A new conceptual framework for social protection, and beyond. International Tax and Public Finance, 8: 529-556.

Jensen, N.M. and Shin, M.J. 2014. Globalization and domestic trade policy preferences: foreign frames and mass support for agriculture subsidies. International Interactions, 40(3): 305-324.

King, D., Gurtner, Y., Firdaus, A., Harwood, S. and Cottrell, A. 2016. Land use planning for disaster risk reduction and climate change adaptation: Operationalizing policy and legislation at local levels. International journal of disaster resilience in the built environment, 7(2): 158-172.

Nesta, L., Vona, F. and Nicolli, F. 2014. Environmental policies, competition and innovation in renewable energy. Journal of Environmental Economics and Management, 67(3): 396-411.

Puntsagdorj, B., Orosoo, D., Huo, X. and Xia, X. 2021. Farmer's perception, agricultural subsidies, and adoption of sustainable agricultural practices: A case from Mongolia. Sustainability, 13(3): 1524.

Purnuş, A. and Bodea, C.N. 2015. Educational simulation in construction project financial risks management. Procedia Engineering, 123: 449-461.

Purushothaman, S., Patil, S. and Francis, I. 2013. Assessing the impact of policy-driven agricultural practices in Karnataka, India. Sustainability Science, 8: 173-185.

Qiao, Y. 2007. Public risk management: development and financing. Journal of Public Budgeting, Accounting & Financial Management, 19(1): 33-55.

Sharma, D. 2007. Agricultural subsidy and trade policies. In Ethics, hunger and globalization: In search of appropriate policies (pp. 263-279). Dordrecht: Springer Netherlands.

3

Data Analytics and Its Role in Optimizing Crop Yields and Resource Utilization

Abhay Deep Gautam

Subject Matter Specialist (GPB), KVK, Chandauli, Uttar Pradesh

Abstract

Data analytics plays a crucial role in optimizing crop yields and resource utilization, particularly in the agriculture sector, where efficiency and productivity are paramount. By harnessing the power of data analytics, farmers and agricultural professionals can make informed decisions that maximize output while minimizing resource wastage. Through data analytics, farmers can analyze various factors such as soil quality, weather patterns, crop health, and historical yields to identify trends and patterns. This information allows them to make precise decisions regarding planting schedules, irrigation, fertilization and pest control, tailored to specific crop and soil conditions. Moreover, data analytics enables predictive modeling, forecasting potential yield outcomes based on different scenarios and inputs. This foresight empowers farmers to proactively address challenges and optimize their strategies for better results. The data-driven insights facilitate the efficient allocation of resources such as water, fertilizers, and pesticides, reducing waste and environmental impact. By optimizing resource utilization, farmers can enhance sustainability practices and mitigate the risks associated with overuse or misuse of inputs. In essence, data analytics serves as a transformative tool in modern agriculture, revolutionizing traditional farming practices by enabling precision agriculture and driving towards more sustainable and productive crop production systems. Its importance in optimizing crop yields and resource utilization cannot be overstated, offering a pathway towards a more resilient and efficient agricultural future.

Keywords: *Resource Utilization, Data Analytics, Modern Agriculture, Crop Health, etc.*

Introduction

In the evolving landscape of agriculture, where feeding a rapidly growing global population is a pressing concern, the role of data analytics has emerged as a beacon of hope, igniting revolutionary transformations in optimizing crop yields and resource utilization. Data analytics, the science of examining raw data to uncover insights and make informed decisions, has transcended conventional farming practices, propelling the industry towards a new era of efficiency, sustainability, and productivity. At the heart of this transformation lies the fusion of cutting-edge technology and age-old agricultural wisdom. By leveraging advanced data analytics techniques, farmers and agricultural experts can decode the intricate tapestry of factors influencing crop growth, from soil composition and weather patterns to pest outbreaks and market demand. This newfound ability to extract actionable insights from vast volumes of data empowers farmers to make precise, data-driven decisions at every stage of the agricultural cycle. One of the most profound impacts of data analytics in agriculture is its role in optimizing crop yields. Through sophisticated data analysis, farmers can discern patterns and trends in crop performance, identifying optimal planting schedules, irrigation regimes, and fertilization strategies tailored to the specific needs of each field. By harnessing the power of predictive modeling, they can anticipate potential challenges and fine-tune their approach to maximize yields while minimizing risks.

Moreover, data analytics enables farmers to unlock the full potential of precision agriculture. By integrating data from sensors, drones, and satellite imagery, they can create detailed maps of their fields, revealing variations in soil fertility, moisture levels, and crop health with unprecedented accuracy. Armed with this information, farmers can deploy resources more efficiently, targeting inputs such as water, fertilizers, and pesticides precisely where they are needed most. This not only optimizes resource utilization but also reduces environmental impact, promoting sustainable farming practices for future generations. The data analytics facilitates agile decision-making in response to changing conditions. By monitoring real-time data streams and leveraging machine learning algorithms, farmers can swiftly adapt their strategies to address emerging threats such as pest infestations or extreme weather events. This dynamic approach to farming empowers farmers to stay one step ahead, mitigating risks and maximizing opportunities in an ever-changing environment. Beyond the farm gate, data analytics is driving innovation across the entire agricultural value chain. From supply chain optimization and market forecasting to crop insurance and financial risk management, data-driven insights are revolutionizing how agricultural businesses operate and thrive in a complex, interconnected world.

Therefore, the importance of data analytics in optimizing crop yields and resource utilization cannot be overstated. As the agricultural industry grapples with the dual challenges of feeding a growing population and mitigating environmental impact, data analytics offers a powerful toolkit for sustainable, resilient, and productive farming practices. By harnessing the power of data, farmers can cultivate a brighter future for agriculture, where abundance coexists with harmony and every harvest brings us closer to a world where no one goes hungry.

Data Collection Techniques

Data collection techniques in the context of agriculture involve gathering a wide range of data types related to crop growth, environmental conditions, soil health, and more. These techniques are essential for generating the data necessary to conduct meaningful analyses and make informed decisions in agricultural settings. Here's an explanation of some key data collection techniques:

1. **Remote Sensing:** Remote sensing techniques involve capturing data about the Earth's surface from a distance, typically using satellites or aircraft. Remote sensing can provide valuable information about vegetation health, soil moisture levels, temperature variations, and land use patterns. Sensors onboard satellites and aircraft capture various wavelengths of light reflected or emitted by the Earth's surface, allowing researchers to analyze these data to derive insights relevant to agriculture.
2. **Sensor Networks:** Sensor networks consist of arrays of physical sensors deployed in agricultural fields to collect real-time data on various environmental parameters. These sensors can measure factors such as soil moisture, temperature, humidity, pH levels, nutrient levels, and weather conditions. Sensor data can be collected at high spatial and temporal resolutions, enabling farmers to monitor crop growth and environmental conditions with precision.
3. **Drones (Unmanned Aerial Vehicles, UAVs):** Drones equipped with sensors and cameras are increasingly used in agriculture for aerial data collection. Drones can capture high-resolution imagery of fields, allowing farmers to monitor crop health, detect pests and diseases, assess irrigation efficiency, and identify areas of stress or nutrient deficiency. Drones offer flexibility and cost-effectiveness compared to traditional aerial surveys and satellite imagery.

4. **Weather Stations:** Weather stations are instruments that measure atmospheric conditions such as temperature, humidity, rainfall, wind speed, and solar radiation. Weather data are crucial for understanding the climatic factors influencing crop growth and development. Automated weather stations installed in agricultural fields provide continuous, real-time weather data, enabling farmers to make timely decisions regarding irrigation, pest management, and other farming practices.
5. **Soil Sampling and Testing:** Soil sampling involves collecting soil samples from different locations within a field and analyzing them to assess soil properties such as nutrient content, pH, organic matter content, and texture. Soil testing provides valuable information for optimizing fertilizer application, selecting appropriate crop varieties, and managing soil health. Advances in soil testing technologies, such as portable soil analyzers and soil scanning devices, have made soil data collection more efficient and accessible to farmers.
6. **Crop and Yield Monitoring:** Monitoring crop growth and yield involves collecting data on plant development, canopy cover, biomass accumulation, and yield metrics such as grain weight or fruit count. This can be done using field measurements, remote sensing techniques, or image analysis algorithms. Crop and yield monitoring data help farmers track the progress of their crops, identify potential problems, and make informed decisions regarding harvest timing and resource allocation.

Overall, these data collection techniques play a crucial role in modern agriculture by providing farmers and agricultural professionals with the information needed to optimize crop yields, enhance resource utilization and make data-driven decisions for sustainable farming practices.

Data Processing and Management

Data processing and management are fundamental aspects of leveraging data effectively in agriculture. These processes involve organizing, cleaning, storing, and analyzing agricultural data to derive actionable insights and support decision-making. Here's an explanation of data processing and management in agriculture:

1. **Data Organization:** The first step in data processing and management is organizing the collected data in a structured manner. This involves categorizing data based on its type, source, and relevance to specific aspects of agricultural operations. For example, data may be organized into categories such as weather data, soil data, crop health data, yield

data and operational data. Organizing data facilitates easy retrieval and analysis when needed.

2. **Data Cleaning:** Agricultural data often contain errors, inconsistencies, and missing values that need to be addressed before analysis. Data cleaning involves identifying and correcting errors, removing duplicate records, filling in missing values through imputation, and ensuring data consistency and integrity. Clean data is essential for accurate analysis and decision-making.

3. **Data Storage Solutions:** Agricultural data come in various formats and sizes, ranging from structured databases to unstructured text and image data. Choosing the appropriate data storage solutions is crucial for efficient data management. This may involve using relational databases, data warehouses, cloud storage platforms, or distributed file systems depending on the volume, velocity, and variety of data being collected.

4. **Data Integration:** Agricultural data often come from diverse sources such as sensors, weather stations, satellite imagery, and farm management software. Integrating data from these disparate sources enables comprehensive analysis and holistic decision-making. Data integration involves harmonizing data formats, resolving inconsistencies, and combining data from multiple sources into a unified dataset for analysis.

5. **Data Security:** Protecting agricultural data from unauthorized access, manipulation, and loss is paramount for maintaining data integrity and confidentiality. Data security measures such as encryption, access controls, user authentication, and data backup systems help safeguard agricultural data against cyber threats, data breaches, and accidental loss.

6. **Data Analysis Tools and Techniques:** Once data is processed and organized, various analytical tools and techniques can be applied to derive insights and extract knowledge from the data. This may involve statistical analysis, machine learning algorithms, data visualization, and predictive modeling. Data analysis enables farmers and agricultural professionals to identify patterns, trends and relationships in the data that can inform decision-making and optimize agricultural practices.

7. **Scalability and Performance:** As agricultural data volumes continue to grow with advancements in data collection technologies, data processing and management systems must be scalable and performant to handle large-scale data processing tasks efficiently. This may involve

parallel processing, distributed computing, and optimization techniques to ensure timely data processing and analysis.

So, the effective data processing and management are essential for harnessing the full potential of agricultural data to optimize crop yields, enhance resource utilization and support sustainable farming practices. By implementing robust data processing and management practices, farmers and agricultural professionals can unlock valuable insights that drive innovation and improvement in agricultural production systems.

Data Analysis and Modeling

Data analysis and modeling are critical components of leveraging agricultural data to optimize crop yields and resource utilization. These processes involve extracting meaningful insights from agricultural data through statistical analysis, machine learning algorithms, and predictive modeling techniques. Here's an explanation of data analysis and modeling in agriculture:

1. **Statistical Analysis:** Statistical analysis techniques are used to explore and summarize agricultural data, identify patterns, and infer relationships between variables. Descriptive statistics such as mean, median, and standard deviation provide insights into the central tendencies and variability of data. Inferential statistics, including hypothesis testing and regression analysis, enable researchers to make predictions and draw conclusions about relationships within the data. Statistical analysis helps farmers understand the factors influencing crop yields, soil health, and environmental conditions.

2. **Machine Learning Algorithms:** Machine learning algorithms are computational models that learn from data to make predictions or decisions without being explicitly programmed. In agriculture, machine learning techniques such as classification, regression, clustering, and neural networks can be applied to analyze complex datasets and extract actionable insights. For example, machine learning models can predict crop yields based on historical data, classify crop diseases from image data, and optimize irrigation schedules to conserve water while maximizing yields.

3. **Predictive Modeling:** Predictive modeling involves building mathematical models based on historical data to forecast future outcomes or trends. In agriculture, predictive models can be used to anticipate crop yields, identify optimal planting dates, and predict pest outbreaks or disease outbreaks. These models incorporate various factors such as

weather data, soil properties, crop management practices, and historical yield data to generate forecasts that help farmers make informed decisions and mitigate risks.

4. **Data Visualization:** Data visualization techniques are used to communicate insights and findings from data analysis and modeling in a visual format. Visualizations such as charts, graphs, maps, and dashboards help farmers and agricultural professionals interpret complex datasets more effectively and identify trends, patterns, and anomalies. Data visualization enhances decision-making by presenting information in a clear, intuitive manner that facilitates understanding and action.

5. **Model Validation and Evaluation:** It is essential to validate and evaluate the performance of data analysis and modeling techniques to ensure their accuracy and reliability. Model validation involves testing the predictive accuracy of models using independent datasets or cross-validation techniques. Evaluation metrics such as accuracy, precision, recall, and F1-score quantify the performance of models and provide insights into their strengths and limitations.

Overall, data analysis and modeling play a crucial role in leveraging agricultural data to optimize crop yields, enhance resource utilization and support sustainable farming practices. By applying statistical analysis, machine learning algorithms, and predictive modeling techniques, farmers and agricultural professionals can make informed decisions that improve productivity, profitability and environmental stewardship in agriculture.

Crop Monitoring and Health Assessment

Crop monitoring and health assessment are essential components of precision agriculture, enabling farmers to monitor the condition of their crops and identify potential issues affecting growth and yield. These processes involve collecting and analyzing data related to crop health, growth stages, nutrient status and pest and disease presence. Here's an explanation of crop monitoring and health assessment in agriculture:

1. **Remote Sensing:** Remote sensing techniques, such as satellite imagery and drones equipped with cameras and sensors, are commonly used for crop monitoring. These technologies capture high-resolution images of agricultural fields, allowing farmers to assess crop health and growth patterns over large areas. Remote sensing data can be used to detect anomalies, such as areas of stress or nutrient deficiency, and monitor changes in crop condition over time.

2. **Spectral Analysis:** Spectral analysis involves analyzing the reflected or emitted light from crops to assess their health and physiological status. Different wavelengths of light are absorbed or reflected by plants in different ways, providing information about factors such as chlorophyll content, water stress, and nutrient uptake. Spectral indices, such as the Normalized Difference Vegetation Index (NDVI) or the Green Normalized Difference Vegetation Index (GNDVI), are commonly used to quantify crop health and monitor changes in vegetation density and vigor.
3. **Image Analysis:** Image analysis techniques are used to extract meaningful information from remote sensing images and aerial photographs. Computer vision algorithms can analyze images to identify crop types, assess canopy cover, detect pests and diseases, and estimate crop yields. Image analysis tools automate the process of crop monitoring and health assessment, enabling farmers to quickly identify areas requiring attention and take corrective actions.
4. **Field Surveys and Sampling:** Field surveys involve visually inspecting crops in the field to assess their health and condition. Farmers and agricultural professionals may conduct regular field surveys to monitor crop growth stages, identify signs of stress or disease, and collect samples for laboratory analysis. Soil and plant tissue samples may be analyzed for nutrient content, pH levels, and the presence of pathogens or pests, providing valuable information for crop management decisions.
5. **Sensor Technologies:** Sensor technologies, such as infrared thermometers, leaf moisture sensors, and chlorophyll meters, provide real-time data on crop health and environmental conditions. These sensors can be deployed in agricultural fields to monitor factors such as temperature, humidity, soil moisture, and nutrient levels. Sensor data help farmers optimize irrigation scheduling, fertilization practices, and pest management strategies to maintain crop health and maximize yields.
6. **Integrated Pest Management (IPM):** Crop monitoring and health assessment are integral components of integrated pest management (IPM) programs, which aim to minimize the use of pesticides while effectively controlling pests and diseases. By monitoring pest populations and disease incidence, farmers can implement targeted interventions such as biological control, cultural practices, and selective pesticide applications to manage pest outbreaks and minimize crop damage.

Overall, crop monitoring and health assessment are essential for optimizing crop yields, minimizing losses and promoting sustainable farming practices. By leveraging remote sensing technologies, image analysis techniques, field surveys and sensor technologies, farmers can proactively monitor crop health, identify potential issues, and implement timely interventions to ensure the success of their crops.

Economic and Environmental Impact Analysis

Economic and environmental impact analysis of data analytics in optimizing crop yields and resource utilization is crucial for understanding the broader implications of adopting data-driven farming practices. By evaluating the economic benefits and environmental consequences, stakeholders can make informed decisions that balance productivity, profitability and sustainability in agriculture. Here's an explanation of the economic and environmental impact analysis:

1. Economic Benefits

a. **Increased Yields:** Data analytics enables farmers to optimize crop management practices, leading to higher yields per hectare. By leveraging insights from data analysis, farmers can make more informed decisions regarding planting schedules, irrigation management, fertilization practices, and pest control measures. Higher yields translate to increased revenue for farmers and improved food security at the regional and global levels.

b. **Cost Savings:** Data-driven farming practices help farmers optimize resource utilization, reducing input costs such as water, fertilizers, and pesticides. By applying inputs more efficiently based on data insights, farmers can minimize waste and lower production costs while maintaining or even increasing yields. Cost savings contribute to improved profitability and economic resilience in agriculture.

c. **Market Access and Competitiveness:** Farmers who adopt data analytics and precision agriculture techniques may gain a competitive edge in the market. By producing higher-quality crops more efficiently, farmers can command premium prices in niche markets or secure contracts with buyers who prioritize sustainability and traceability. Data-driven farming practices enhance market access and competitiveness for agricultural producers.

2. Environmental Impact

a. **Resource Efficiency:** Data analytics helps optimize resource utilization in agriculture, leading to reduced environmental impact. By applying inputs such as water, fertilizers, and pesticides more precisely based on crop needs and soil conditions, farmers can minimize nutrient runoff, soil erosion, and water pollution. Resource-efficient farming practices promote environmental sustainability and protect natural resources for future generations.

b. **Reduced Chemical Use:** Precision agriculture techniques enabled by data analytics allow farmers to target pest and disease management more effectively, reducing the need for chemical pesticides. By monitoring crop health and pest populations in real-time, farmers can implement integrated pest management (IPM) strategies that rely on biological control, cultural practices, and natural predators rather than synthetic chemicals. Reduced chemical use mitigates environmental pollution and preserves biodiversity in agricultural ecosystems.

c. **Carbon Footprint Reduction:** Adopting data-driven farming practices can contribute to reducing the carbon footprint of agriculture. By optimizing inputs and reducing waste, farmers can lower greenhouse gas emissions associated with agricultural activities such as fertilizer production, fuel consumption, and land use changes. Sustainable farming practices enabled by data analytics help mitigate climate change impacts and enhance resilience in agriculture.

So, the economic and environmental impact analysis of data analytics in agriculture underscores the importance of balancing productivity with sustainability. By maximizing economic benefits while minimizing environmental costs, data-driven farming practices can contribute to a more resilient, equitable and environmentally sustainable food system. Policymakers, researchers, and stakeholders play a critical role in promoting the adoption of data analytics in agriculture and ensuring that its benefits are realized while mitigating potential drawbacks.

Integration with IoT and Emerging Technologies

The integration of data analytics with the Internet of Things (IoT) and other emerging technologies holds immense potential for optimizing crop yields and resource utilization in agriculture. By leveraging IoT devices, sensors, and advanced analytics, farmers can collect real-time data on environmental conditions, crop health, and resource usage, enabling more informed decision-making and precise management practices. Here's a detailed explanation of how integration with IoT and emerging technologies enhances data analytics in agriculture:

1. **Real-Time Monitoring:** IoT devices and sensors deployed in agricultural fields continuously collect data on soil moisture levels, temperature, humidity, rainfall, and other environmental variables. This real-time monitoring provides farmers with up-to-date information on crop conditions, allowing them to respond quickly to changes and optimize resource allocation accordingly. For example, farmers can adjust irrigation schedules based on soil moisture data or implement pest control measures in response to early signs of infestation detected by sensors.
2. **Precision Agriculture:** Integration with IoT enables precision agriculture practices that tailor farming inputs and operations to specific field conditions and crop needs. By combining data from sensors with advanced analytics, farmers can create detailed maps of their fields, identifying areas with varying soil properties, moisture levels, and crop health. This information allows for targeted application of water, fertilizers, and pesticides, minimizing waste and maximizing yields while reducing environmental impact.
3. **Remote Monitoring and Management:** IoT-enabled devices such as drones and remote-controlled machinery allow farmers to remotely monitor and manage their fields from anywhere. Drones equipped with cameras and sensors capture aerial imagery and collect data on crop health, pest infestations, and field conditions, providing valuable insights for decision-making. Remote-controlled machinery equipped with IoT sensors can perform tasks such as precision seeding, spraying, and harvesting, optimizing labor efficiency and reducing operational costs.
4. **Data Integration and Analysis:** Integration with IoT expands the scope and granularity of data available for analysis, enabling more comprehensive insights into agricultural systems. By integrating data from IoT devices with weather forecasts, satellite imagery, soil maps, and historical data, farmers can gain a holistic view of their operations and identify trends, patterns, and correlations that inform decision-making. Advanced analytics techniques such as machine learning and predictive modeling can process large volumes of IoT data to generate actionable insights and optimize crop management strategies.

5. **Smart Irrigation and Resource Management:** IoT-enabled irrigation systems use sensors to monitor soil moisture levels and weather conditions in real-time, adjusting water application rates and schedules accordingly. Smart irrigation reduces water waste, improves water-use efficiency, and minimizes environmental impact by avoiding overwatering and runoff. Similarly, IoT-based systems for monitoring nutrient levels, pest populations, and crop health enable precise resource management, optimizing fertilizer and pesticide use while minimizing environmental harm.

The integration with IoT and emerging technologies enhances data analytics capabilities in agriculture, enabling farmers to optimize crop yields and resource utilization with unprecedented precision and efficiency. By leveraging real-time data, advanced analytics, and smart management practices, farmers can improve productivity, profitability and sustainability in modern agriculture while minimizing environmental impact. Continued innovation and adoption of IoT technologies hold the promise of transforming agriculture into a more resilient, resource-efficient, and environmentally sustainable industry.

Policy and Regulatory Considerations

Policy and regulatory considerations play a crucial role in shaping the adoption and implementation of data analytics for optimizing crop yields and resource utilization in agriculture. While data analytics offers significant potential for improving productivity and sustainability in farming, policymakers must address various challenges related to data privacy, security, ownership, and access. Here's a detailed explanation of policy and regulatory considerations:

1. **Data Privacy and Security:** Agricultural data collected through data analytics platforms may contain sensitive information about farm operations, crop yields, and resource utilization. To protect farmers' privacy and prevent unauthorized access or misuse of data, policymakers must establish robust data privacy and security regulations. These regulations may include requirements for data encryption, anonymization and access controls and data breach notification procedures to safeguard agricultural data against cyber threats and unauthorized access.

2. **Data Ownership and Control:** Clarifying ownership rights and control over agricultural data is essential to ensure that farmers retain ownership of their data and have control over how it is collected, used, and shared. Policymakers may need to develop regulations and contractual frameworks that clarify data ownership rights between farmers, agricultural service providers, technology vendors, and other

stakeholders involved in data analytics initiatives. Clear guidelines on data ownership and control help foster trust and transparency in data-driven farming practices.

3. **Data Sharing and Interoperability:** Promoting data sharing and interoperability among agricultural stakeholders is critical for maximizing the benefits of data analytics in agriculture. Policymakers can encourage data sharing through incentives, standards development, and public-private partnerships that facilitate collaboration and interoperability among data platforms and systems. Open data initiatives and data sharing agreements may help overcome barriers to data access and promote innovation in agricultural data analytics.
4. **Ethical and Responsible Data Use:** Policymakers must address ethical considerations related to the use of agricultural data, ensuring that data analytics initiatives adhere to ethical principles and respect farmers' rights and interests. Regulations may be needed to establish guidelines for responsible data use, including principles of fairness, transparency, accountability, and non-discrimination. Ethical frameworks for data analytics in agriculture can help mitigate risks such as algorithmic bias, privacy violations, and misuse of data for commercial or political purposes.
5. **Regulatory Compliance and Standards:** Policymakers may need to develop regulatory frameworks and standards for data analytics in agriculture to ensure compliance with legal requirements and industry best practices. Regulations may cover aspects such as data collection, storage, processing, sharing, and use, as well as standards for data formats, metadata, and interoperability. Compliance with regulatory requirements helps mitigate legal risks and ensures that data analytics initiatives adhere to relevant laws and regulations governing agriculture and data privacy.

Conclusion

In conclusion, data analytics stands as a transformative force in agriculture, revolutionizing traditional farming practices and paving the way for a more sustainable and productive future. By harnessing the power of data, farmers can optimize crop yields and resource utilization with unprecedented precision, efficiency, and sustainability. Through data analytics, farmers can make informed decisions at every stage of the agricultural cycle, from planting and irrigation to pest control and harvest planning. By leveraging advanced analytics techniques, such as machine learning and predictive modeling, farmers can anticipate challenges, identify opportunities and optimize their strategies to maximize productivity while minimizing environmental impact. The integration with emerging technologies such as IoT devices, drones and remote sensing platforms further enhances the capabilities of data analytics in agriculture, enabling real-time monitoring, smart management practices and data-driven decision-making. However, realizing the full potential of data analytics in agriculture requires addressing various challenges, including data privacy, security, ownership and regulatory compliance. Collaborative efforts among policymakers, farmers, industry stakeholders and researchers are essential to develop regulatory frameworks that support responsible data use and foster innovation in data-driven farming practices. Overall, data analytics holds the promise of transforming agriculture into a more resilient, efficient and sustainable industry, ensuring food security and environmental stewardship for generations to come.

References

Balaselvakumar, S. and Saravanan, S. 2006. Remote sensing techniques for agriculture survey.

Belcore, E., Angeli, S., Colucci, E., Musci, M.A. and Aicardi, I. 2021. Precision agriculture workflow, from data collection to data management using FOSS tools: an application in northern Italy vineyard. ISPRS International Journal of Geo-Information, 10(4): 236.

Blandford, D. 2007. Information deficiencies in agricultural policy design, implementation and monitoring.

Dewi, C. and Chen, R.C. 2020. Decision making based on IoT data collection for precision agriculture. Intelligent Information and Database Systems: Recent Developments, 11: 31-42.

Ehlers, M.H., Huber, R. and Finger, R. 2021. Agricultural policy in the era of digitalization. Food Policy, 100: 102019.

Inyim, P., Rivera, J. and Zhu, Y. 2015. Integration of building information modeling and economic and environmental impact analysis to support sustainable building design. Journal of Management in Engineering, 31(1): 4014002.

Judijanto, L. 2023. Analysis of Weather Prediction, Resource Management, and Land Optimization on the Application of Big Data Analytics in Agricultural Land Utilization in Agrarian Areas of West Java. West Science Nature and Technology, 1(02): 81-90.

Koshariya, A.K., Rameshkumar, P.M., Balaji, P., Cavaliere, L.P.L., Dornadula, V.H.R. and Singh, B. 2024. Data-Driven Insights for Agricultural Management: Leveraging Industry 4.0 Technologies for Improved Crop Yields and Resource Optimization. In Robotics and Automation in Industry 4.0: 260-274. CRC Press.

Ruben, R., Moll, H. and Kuyvenhoven, A. 1998. Integrating agricultural research and policy analysis: Analytical framework and policy applications for bio-economic modelling. Agricultural systems, 58(3): 331-349.

Sagar, B.M. and Cauvery, N.K. 2018. Agriculture data analytics in crop yield estimation: a critical review. Indonesian Journal of Electrical Engineering and Computer Science, 12(3): 1087-1093.

Van der Werf, H.M. and Petit, J. 2002. Evaluation of the environmental impact of agriculture at the farm level: a comparison and analysis of 12 indicator-based methods. Agriculture, Ecosystems & Environment, 93(1-3): 131-145.

4

Climate Change and Its Impact on Agriculture: Global and National Level Initiatives and Plans to Support Climate Change

[1]Shivendra Mishra, [2]Abdhesh Kumar, [3]Mani Prakash Shukla and [3]Pawan Kumar Gupta

[1]Dairy Extension Division
ICAR-National Dairy Research Institute, Karnal (Haryana)
[2]Department of Agricultural Extension Education, Banda University of Agriculture and Technology, Banda (Uttar Pradesh)
[3]Department of Agricultural Extension and Education, Chandra Shekhar Azad University of Agriculture Technology, Kanpur (Uttar Pradesh)

Abstract

A shift in the long-term weather patterns that define different parts of the world is referred to as global climate change. The short-term (daily) variations in a region's temperature, wind, and/or precipitation are referred to as the "weather" (Merritts et al. 1998). Long-term climate change may have a variety of effects on agriculture, including changes in crop productivity, growth rates, photosynthesis and transpiration rates, moisture availability, and crop quantity and quality. Global food production is expected to be directly impacted by climate change. A rise in the average seasonal temperature can shorten the growing season for many crops, which will lower production. Warming will have an immediate effect on yields in regions where temperatures are already near the physiological maximum for crops (IPCC, 2007). Through their effects on plant physiology, drivers of climate change can also directly affect food production through changes in air composition. Special agricultural measures are necessary to battle the severe repercussions of agriculture's contribution to climate change and the negative effects of climate change on agriculture. These implications are expected to have a significant influence on food production and may even jeopardize food security.

Keywords: *Climate, Weather, Global, Production, agriculture, etc.*

Introduction

Any notable long-term alteration to the anticipated patterns of the average weather in an area (or the entire Earth) over an extended period of time is referred to as climate change. It concerns anomalous fluctuations in the climate and how these fluctuations affect other regions of the planet. Tens, hundreds, or even millions of years may pass before these changes occur. However, as a result of increased anthropogenic activity including industrialization, urbanization, deforestation, agriculture, and altered land use patterns, greenhouse gas emissions grow and accelerate the rate of climate change. Increases in temperature, variations in precipitation, and an increase in atmospheric CO_2 concentrations are all part of climate change scenarios. There are three potential ways that agriculture might benefit from the greenhouse effect. First, the pace at which weeds and agricultural plants develop can be directly impacted by rising atmospheric CO_2 concentrations. Second, temperature, precipitation, and sunlight variations brought on by CO_2 may have an impact on the production of plants and animals. Finally, because of flooding and rising groundwater salinity in coastal regions, rising sea levels may result in the loss of cropland. The earth's climate is mostly shaped by a natural mechanism called the greenhouse effect. It creates the area close to the earth's surface that is comparatively warm and friendly, allowing living forms, including humans, to evolve and flourish. Global warming, however, has resulted from an increase in greenhouse gases (GHGs) caused by human activity. These gases include carbon dioxide (CO_2), water vapor (H_2O), methane (CH_4), nitrous oxide (N_2O), hydrofluorocarbons (HFCs), perfluorocarbons (PFCs), and sulfur hexafluoride (SF6). The average global surface temperature having grown by 0.74°C since the late 19th Century and is anticipated to climb by 1.4°C - 5.8°C by 2100 AD with major regional differences (IPCC, 2007). Between 1750 and 2012, there was a rise in the atmospheric content of CO_2 (280 ppm to 395 ppm), CH_4 (715 ppb to 1882 ppb), and N_2O (227 ppb to 323 ppb). The CO_2, CH_4, and N_2O gases have a Global Warming Potential (GWP) of 1, 25, and 310, respectively.

In the northern regions of India, the warming can be more noticeable. As the climate changes, it is anticipated that the extremes of maximum and lowest temperatures will rise. Some regions may see more precipitation, while others may experience dry spells. With the exception of Tamil Nadu in the south and Punjab and Rajasthan in the west, where average summer monsoon rainfall throughout all states of India has increased by 20%. Though the number of rainy days may decrease in certain regions of India (like MP), most regions should see an increase in intensity (like the North East). The gross per capita availability of water in India is expected to decrease from 1820 m^3/year in

2001 to a maximum of 1140 m^3/year by 2050. Summertime temperatures in the Indian Ocean that surpass the thermal thresholds recorded for the previous two decades may soon be experienced by corals in that region. Beginning in 2050, annual coral bleaching will very certainly occur. At present, the districts of Odisha's Jagatsinghpur and Kendrapara, Tamilnadu's Nellore and Nagapattinam, and Gujarat's Junagadh and Porabandar districts are particularly susceptible to the effects of cyclones occurring more frequently and with greater severity (NATCOM, 2004). The mean sea level along the Indian coast has been seen to be increasing at a rate of around 1.0 mm/year during a 100-year period. On the other hand, current evidence points to a 2.5 mm/year rise in sea level along the Indian coastline. By the middle of this century, the sea surface temperature next to India is expected to rise by around 1.5–2.0°C, and by the end of the century, by approximately 2.5–3.5°C. An estimated 7.1 million people in India are expected to be displaced by a 1 meter increase in sea level, and 4200 km of highways and 5764 sq km of land will be destroyed (NATCOM, 2004). The kinds of almost 50% of India's forests are predicted to change, which would have a negative effect on the biodiversity that goes along with it, the dynamics of the local climate, and the livelihoods that depend on forest products. The majority of the forest biomass in India appears to be quite sensitive to the anticipated change in climate, even in the relatively short time span of around 50 years. Furthermore, it is predicted that the kinds of forests in 77% and 68% of India's wooded grids would probably change by 2085.

Impact of climate change on India's agriculture

Since ancient times, India's agriculture has been mostly depended on the monsoon. A shift in the monsoon pattern has a significant impact on agriculture. Indian agriculture is being impacted even by the rising temperatures. These pre-monsoon variations in the Indo-Gangetic Plain will mainly impact the wheat crop (>0.5°C increase in time slice 2010-2039; IPCC 2007). The losses in rice output during severe droughts (which occur around once every five years) in the states of Jharkhand, Odisha, and Chhattisgarh alone amount to an estimated $800 million, or almost 40% of overall production. Crop yields of rice, wheat, legumes, and oilseeds increase by 10–20% as CO^2 levels rise to 550 ppm. Temperature increases of 1°C may result in 3-7% lower yields of potatoes, groundnuts, mustard, soybeans, and wheat. At higher temperatures, much larger losses. Most crops' productivity will only slightly decline by 2020, but it will rise by 10–40% by 2100 as a result of rising temperatures, more erratic rainfall, and less irrigation water. Nearly 60% of farmland is used for rain-fed or unirrigated crops, which will be the main targets of climate change effects. A temperature rise by 0.5°C in winter temperature is projected to reduce rain fed wheat yield by 0.45 tonnes per hectare in India. In India, it is predicted that a 0.5°C increase in winter temperatures will result in a 0.45 tons per hectare decrease in rain-fed wheat output. There may have been some improvement in the yields of coconut, rabi, chickpeas, sorghum, and millets in the west coast. Reduced frost damage has resulted in less loss of vegetables, potatoes, and mustard in northwest India. A rise in floods and droughts is probably going to increase the unpredictability of production. According to recent research conducted at the Indian Agricultural Research Institute, every degree Celsius that the temperature rises during the growing season might result in a loss of 4–5 million tons of wheat output in the future. The yield of rice is predicted to decrease by around one tonne per hectare for every 2°C rise in temperature. A 2°C increase in temperature in Rajasthan was predicted to cause a 10%–15% decrease in pearl millet yield. A 3°C increase at the highest and a 3.5°C increase at the minimum temperature would result in a 5% decrease in M.P. soybean yields by comparison to 1998. The coastal districts of Gujarat and Maharashtra would see the greatest impact on agriculture due to the susceptibility of productive areas to salinization and flooding.

Impact of climate change on world's agriculture

Global food production is expected to be directly impacted by climate change. A rise in the average seasonal temperature can shorten the growing season for many crops, which will lower their yield in the end. Warming will have an immediate effect on yields in regions where temperatures are already near the physiological maximum for crops (IPCC, 2007). Global warming will cause a significant drop in world agriculture this century. By 2080, it is predicted that global agricultural production would have decreased by 3 to 16 percent. It is expected that agricultural production in developing nations—many of which now have average temperatures that are close to or beyond crop tolerance levels—will decrease by 10 to 25% on average by the year 2080. Rich nations, whose average temperatures are normally lower, may have a considerably milder or even positive average effect, with output potentially declining by 6% or increasing by 8%. Significant reductions are also faced by individual emerging nations. For example, India may experience a 30–40% decline.

Global and National Level Initiatives and Plans to Support Climate Change Adaptation

1. Global Level Initiatives and Plans

Several global organizations play a significant role in addressing climate change adaptation by conducting research, setting standards, facilitating cooperation and implementing projects. Here is a list of some prominent global organizations working on climate change:

1.1 **United Nations Framework Convention on Climate Change (UNFCCC)**

Established in 1992 against rising concerns about global climate change, the United Nations Framework Convention on Climate Change (UNFCCC) serves as a comprehensive international framework. Headquartered in Bonn, Germany, it involves 198 member countries and provides objectives, principles and commitments for addressing climate change. India is a signatory, having ratified it in 1993. A major milestone for the UNFCCC occurred with the adoption of the Paris Agreement in 2015. This landmark global treaty aims to fortify the collective response to climate change. The agreement sets a crucial goal to limit global warming to well below 2°C, with an ambitious target of 1.5°C. It establishes a structured framework for countries to submit and update their emission reduction targets, encourages adaptation measures and facilitates financial and technological support (Handbook United Nations Framework Convention on Climate Change, 2006). The Nationally Determined Contributions (NDCs) became a central component of the Paris Agreement, and the agreement itself entered into force on November 4, 2016. The NDCs are integral to the Paris Agreement's objective of limiting global temperature increases and adapting to the impacts of climate change.

2.1.1 Purpose of NDCs

1. NDCs within the Paris Agreement are crucial for each country's efforts to adapt to the impacts of climate change.
2. The Paris Agreement mandates each Party to formulate, communicate and maintain successive NDCs, expressing their intentions to adapt to the challenges posed by climate change.

2.1.2 Adaptation Measures

1. Parties are required to implement domestic adaptation measures with the objective of fulfilling the adaptation goals outlined in their contributions.
2. The emphasis is on addressing the adverse effects of climate change, particularly for vulnerable countries, including the least developed countries (LDCs) and Small Island Developing States (SIDS).

2.1.3 Global Goals for Adaptation

1. The global goals of the Paris Agreement include recognizing the urgency of adaptation to climate change.
2. Parties commit to supporting result-based measures for the development of approaches at all levels on vulnerability and adaptation, including capacity-building for integrating adaptation concerns into sustainable development strategies.

2.1.4 Submission and Review of NDCs

1. NDCs, with a focus on adaptation, are submitted every five years to the UNFCCC secretariat.
2. The Paris Agreement encourages a continuous increase in ambition for adaptation measures, ensuring that each successive NDC represents progress compared to the previous one.
3. The first round of NDCs was due by 2020 and subsequent submissions occur every five years thereafter (e.g., 2025, 2030).

2.1.5 Flexibility for Adjustment

1. Parties have the flexibility to adjust their existing NDCs at any time to enhance their level of ambition in adaptation measures.
2. This adjustment provision allows countries to respond to changing circumstances and strive for more ambitious climate adaptation actions (*Nationally Determined Contributions (NDCs) | UNFCCC*, n.d.).

2.2 Intergovernmental Panel on Climate Change (IPCC)

The IPCC is a scientific body established by the United Nations (UN) in 1988. The IPCC creates thorough Assessment Reports that detail the current level of scientific, technical and socioeconomic understanding on climate change, its effects, potential dangers in the future and strategies for slowing down its rate of occurrence. It also releases Methodology Reports, which offer instructions for creating greenhouse gas inventories and Special Reports on subjects decided upon by its member nations. The most recent report is the Sixth Assessment Report, which includes a Synthesis Report and three contributions from Working Groups. The Working Group I (Assesses the physical science basis of climate change), Working Group II (Assesses impacts, adaptation and vulnerability related to climate change) and Working Group III (Assesses mitigation options and strategies to reduce greenhouse gas emissions) completed their contributions in August 2021, February 2022, April 2022, and March 2023, respectively, and the Synthesis Report was completed in August 2023(IPCC — Intergovernmental Panel on Climate Change, 2023).

2.3 Green Climate Fund (GCF)

The Green Climate Fund (GCF) was established in 2010 under the Cancun Agreements and is a financial mechanism under the United Nations Framework Convention on Climate Change (UNFCCC). Its primary mission is to enhance global efforts to respond to climate change by mobilizing substantial funding for low-emission and climate-resilient development. GCF operates with a balanced governance structure, guided by a Board of twenty-four members from both developed and developing countries. This consensus-based decision-making body oversees the Fund's activities, and a permanent Secretariat, established in 2013, supports its work from Songdo, Republic of Korea.

The Fund's key objectives include achieving a 50/50 balance in funding allocation between mitigation and adaptation projects. Notably, over 50% of adaptation funding is directed to the most vulnerable countries, including Least Developed Countries (LDCs), Small Islands Developing States (SIDS) and African States. GCF aims to attract private investment by de-risking high-impact adaptation projects in developing countries. In terms of financing, GCF successfully raised over USD 10.3 billion in pledges from 35 countries during its initial resource mobilization in 2014. These funds, covering the period from 2015 to 2018, provide the means for GCF to embark on its mitigation and adaptation activities. Looking ahead, GCF faces the challenge of delivering on its ambitious mandate, contributing to a real paradigm shift in building climate resilience and supporting the transition to low-emission economies. The Fund's initiatives and plans align with international climate goals, emphasizing the importance of country-driven approaches, financial innovation and a balanced focus on mitigation and adaptation efforts (Manzanares, 2017).

2.4 United Nations Environment Programme (UNEP)

The United Nations Environment Programme (UNEP) was established in 1972. Its fundamental function is to provide leadership and encourage global collaboration in addressing environmental issues. UNEP works to develop international environmental policies, conduct scientific research, build capacity for sustainable environmental management and catalyze partnerships among governments, the private sector and civil society to promote the wise use of natural resources and protect the global environment.

The Programme of Research on Climate Change Vulnerability, Impacts, and Adaptation (PROVIA) is an initiative that aims to address the challenge of providing relevant information to policymakers on vulnerability, impacts and adaptation (VIA) in the context of climate change. PROVIA operates in a coherent and coordinated manner to harmonize, mobilize and communicate the growing knowledge base on VIA to relevant audiences. PROVIA is associated with the United Nations Environment Programme (UNEP). UNEP is a key partner in supporting and coordinating various global initiatives related to environmental issues, including climate change. PROVIA's activities are aligned with UNEP's broader goals in addressing climate change impacts and supporting adaptation efforts. The initiative collaborates with experts, researchers and policymakers to develop research priorities, gather knowledge, and facilitate communication to inform decision-making processes related to climate change vulnerability and adaptation (Rosenzweig *et al.*, 2013).

2.5 World Bank

The World Bank is a global financial institution established on July 1, 1944, focused on reducing poverty and supporting sustainable development. In the realm of climate change, it plays a pivotal role by providing climate finance, funding projects for mitigation and adaptation, offering technical assistance and sharing knowledge to help countries integrate climate considerations into their development strategies. The World Bank collaborates with various stakeholders to address climate challenges and promote resilient and sustainable development.

The World Bank Group has recently introduced a comprehensive Climate Change Action Plan that will shape its endeavours from 2021 to 2025. This strategic initiative is designed to escalate climate finance, with a primary focus on diminishing emissions, fortifying climate change adaptation measures and ensuring that financial activities are in harmony with the objectives outlined in the Paris Agreement. The plan underscores the World Bank's commitment to addressing the challenges posed by climate change and signals a proactive approach to supporting adaptation efforts worldwide. Through this Action Plan, the World Bank aims to play a pivotal role in advancing sustainable development by allocating resources to projects that enhance adaptive capacity and contribute to global climate resilience (*World Bank Group Climate Change Action Plan (2021-2025) Infographic*, n.d.).

2.6 Global Green Growth Institute (GGGI)

The Global Green Growth Institute (GGGI) officially launched in 2012, driven by South Korean President Lee Myung-bak's vision of reshaping economic growth. With a focus on green growth, the GGGI operates with a "me first" approach, urging countries to proactively execute policies for sustainable economic development. The institute, having evolved from a small nonprofit to an international organization, is actively engaged in ten countries, providing technical and policy advice to integrate green growth into their economic development agendas. Core to its initiatives is advising countries on forming Green Growth Plans (GGPs), which involve assessing economic goals, identifying overlaps with environmental benefits, and facilitating investment analysis. The GGGI's unique whole-of-government approach seeks to integrate green growth into national economic development agendas, differentiating it from traditional environmental projects. As it expands its operational presence, the GGGI's success in proving the feasibility of green growth and addressing economic and environmental challenges will be crucial for achieving its transformative goals (O'Donnell, 2012).

3. National Level Initiatives and Plans

3.1 National Action Plan on Climate Change (NAPCC)

The Prime Minister unveiled the National Action Plan on Climate Change (NAPCC) on June 30, 2008, presenting a comprehensive national strategy to adapt to climate change and enhance India's development path's ecological sustainability. Emphasizing the crucial role of maintaining a high growth rate, the plan aims to improve living standards for the majority of Indians while reducing vulnerability to climate change impacts. The NAPCC is structured around eight National Missions, including the National Solar Mission, National Mission for Enhanced Energy Efficiency, National Mission on Sustainable Habitat, National Water Mission, National Mission for Sustaining the Himalayan Eco-system, National Mission for a Green India, National Mission for Sustainable Agriculture and National Mission on Strategic Knowledge for Climate Change. These missions collectively concentrate on fostering awareness of climate change, promoting adaptation and mitigation measures, enhancing energy efficiency, and conserving natural resources (*National Action Plan on Climate Change (NAPCC)*, 2021).

The following The National Action Plan on Climate Change (NAPCC), launched in 2008, comprises eight missions, each contributing to climate change adaptation. The Committee on Estimates' recommendations highlight the importance of effective planning and fund utilization for these missions to significantly contribute to climate change adaptation in India as follows (*Committee Reports*, n.d.):

1. National Solar Mission: Aiming for 1,00,000 MW of solar power by 2021-22, supporting clean energy for climate resilience.
2. National Mission on Enhanced Energy Efficiency: Emphasizing energy efficiency to reduce the carbon footprint and aid climate adaptation.
3. National Water Mission: Focused on comprehensive water management and assessing climate change impacts on water resources.
4. National Mission on Sustainable Habitat: Promoting urban sustainability, indirectly contributing to climate resilience through efficient energy and waste management.

5. National Mission for Sustainable Agriculture: Addressing various aspects of agriculture, including income security for farmers, aligning with climate change adaptation.

3.2 National Adaptation Fund for Climate Change (NAFCC)

The National Adaptation Fund for Climate Change (NAFCC), established in 2015-16 as a Central Sector Scheme, operates in project mode to support adaptation activities mitigating climate change impacts. Eligible sectors for funding include agriculture, animal husbandry, water, forestry, tourism, and more. NABARD serves as the National Implementing Entity. The objectives include financing adaptation projects aligned with National Action Plan on Climate Change (NAPCC) and State Action on Climate Change (SAPCCs), climate scenario preparation, vulnerability assessment, capacity building and knowledge network development. The outcome framework targets reduced climate risks, cross-sectoral benefits, enhanced community awareness, strengthened capacity, increased adaptive capacity, improved policies and diversified livelihoods. Project proposals are prepared in consultation with NABARD, requiring approval from the State Steering Committee on Climate Change for NAFCC consideration. States have the discretion to engage organizations for project preparation as needed (*National Adaptation Fund for Climate Change — Vikaspedia*, n.d.).

Project sanctioned under National Adaptation Fund for Climate Change for the state of Haryana is "Scaling-up Climate Resilient Agriculture Practices towards Climate Smart Villages (CSVs) in Haryana". The objective of the project is to improve the adaptive capacity of rural community to climate change through enhancing the portfolios of climate resilient agriculture interventions in targeted villages of Haryana. The project would be undertaken in 250 villages in 10 districts of Haryana viz, Yamunanagar, Ambala, Kurukshetra, Karnal, Jind, Kaithal, Panipat, Sonipat, Sirsa and Fatehabad covering 75,000 families. Department of Agriculture, Government of Haryana is executing the project with an outlay of Rs. 22.10 Crores (NABARD - National Bank for Agriculture and Rural Development, n.d.).

3.3 National Innovations on Climate Resilient Agriculture (NICRA)

Launched in February 2011 by the Indian Council of Agricultural Research (ICAR), the National Innovations on Climate Resilient Agriculture (NICRA) is a network project designed to fortify Indian agriculture against climate change. With an initial outlay of Rs. 350 crores for the XI Plan, NICRA addresses the pressing need for climate-resilient agriculture through strategic research, technology demonstration, capacity building, and sponsored/competitive grants.

Objectives

1. Enhancing Resilience: NICRA seeks to enhance the resilience of Indian agriculture, covering crops, livestock, fisheries, and natural resource management.
2. Demonstrating Technologies: The project demonstrates site-specific technology packages on farmers' fields to adapt to current climate risks.
3. Capacity Building: NICRA aims to enhance the capacity of scientists and stakeholders in climate-resilient agricultural research and its practical application.

Components

1. **Strategic Research:** The research spans crops, horticulture, livestock, natural resource management, and fisheries, focusing on vulnerability assessment, agro-advisories, evolving varieties, greenhouse gas monitoring, water and nutrient use efficiency, and more.
2. **Technology Demonstration:** Implemented in 100 vulnerable districts, this component introduces proven technologies, from in-situ moisture conservation to introducing climate-tolerant varieties, involving over one lakh farm families.
3. **Sponsored and Competitive Grants:** This component addresses critical research gaps, including impact on pollinators, fisheries, hailstorm management, hill ecosystems, and socio-economic aspects of climate change.
4. **Capacity Building:** NICRA conducts training programs for young scientists on simulation modelling, high-throughput phenotyping, greenhouse gas measurement, and awareness programs covering 50,000 farmers (*Welcome To NICRA*, n.d.).

3.4 National Adaptation Programmes of Action (NAPAs)

In response to the distinctive climate change vulnerabilities faced by least developed countries (LDCs), the Conference of the Parties (COP) initiated the Least Developed Countries (LDC) work programme in 2001. A pivotal component of this program is the establishment of National Adaptation Programmes of Action (NAPAs, 2005), aimed at aiding LDCs in addressing the specific challenges associated with climate change. This comprehensive framework also introduced the Least Developed Countries Fund (LDCF) to financially support the formulation and execution of NAPAs. Additionally, the creation of the LDC Expert Group (LEG) was pivotal, providing crucial technical support and guidance to empower these nations in navigating the complexities of climate change adaptation. National Adaptation Programmes of Action (NAPAs) submitted by 41 Least Developed Countries (LDCs) to the United Nations Framework Convention on Climate Change (UNFCCC), examining their integration with national development strategies. The analysis reveals that, among the 41 NAPAs, 37 acknowledge the link between high and rapid population growth and climate change. Additionally, six NAPAs explicitly advocate for measures such as slowing population growth and investing in reproductive health/family planning (RH/FP) as priority adaptation actions. However, the study identifies structural challenges hindering better alignment between adaptation planning and national development and provides recommendations for more effective long-term adaptation strategies. The intersection of population dynamics with climate change and national development agendas is explored, emphasizing the need for cohesive planning that addresses these interconnected challenges (Hardee & Mutunga, 2010).

Conclusion

Climate change, the outcome of the "Global Warming" has now started showing its impacts worldwide. Climate is the primary determinant of agricultural productivity which directly impact on food production across the globe. Agriculture sector is the most sensitive sector to the climate changes because the climate of a region/country determines the nature and characteristics of vegetation and crops. Food production systems are extremely sensitive to climate changes like changes in temperature and precipitation, which may lead to outbreaks of pests and diseases thereby reducing harvest ultimately affecting the food security of the country. Coping with the impact of climate change on agriculture will require careful management of resources like soil, water and biodiversity. To cope with the impact of climate change on agriculture and food production, India will need to act at the global, regional, national and local levels. The global response to climate change adaptation involves collaborative efforts at various levels. Global initiatives led by organizations like the UNFCCC, IPCC, and GCF set ambitious goals, while countries like India implement comprehensive national plans. Despite challenges, opportunities arise from technological advancements, international collaboration, private sector innovation and community empowerment. The interconnected efforts outlined emphasize the need for sustained commitment, increased collaboration and a transformative shift towards proactive, location-specific adaptation strategies. Integrating adaptation seamlessly into development plans paves the way for a more resilient and sustainable future amidst a changing climate.

References

Hardee, K., & Mutunga, C. 2010. Strengthening the link between climate change adaptation and national development plans: Lessons from the case of population in National Adaptation Programmes of Action (NAPAs). Insectes Sociaux, 57(2), 113–126. https://doi.org/10.1007/S11027-009-9208-3/TABLES/2

http://en.wikipedia.org/wiki/Agriculture_in_India

http://www.ipcc.ch/pdf/assessmentreport

IPCC — Intergovernmental Panel on Climate Change. (n.d.). Retrieved 28 January 2024, from https://www.ipcc.ch

Manzanares, F. J. 2017. The Green Climate Fund – a beacon for climate change action. Asian Journal of Sustainability and Social Responsibility 2016 2:1, 2(1), 1–5. https://doi.org/10.1186/S41180-016-0012-1

NABARD - National Bank For Agriculture And Rural Development. (n.d.). Retrieved 28 January 2024, from https://www.nabard.org/content.aspx?id=585

National Action Plan on Climate Change (NAPCC). 2021. https://dst.gov.in/climate-change-programme.

National Adaptation Fund for Climate Change — Vikaspedia. (n.d.). Retrieved 28 January 2024, from https://vikaspedia.in/energy/environment/climate-change/national-adaptation-fund-for-climate-change.

Nationally Determined Contributions (NDCs) UNFCCC. (n.d.). Retrieved 28 January 2024, from https://unfccc.int/process-and-meetings/the-paris-agreement/nationally-determined-contributions-ndcs

Rosenzweig, C., Horton, R. M., Rosenzweig, C., & Horton, R. M. (2013). Research Priorities on Vulnerability, Impacts and Adaptation: Responding to the Climate Change Challenge. AGUFM, 2013, ED12B-05. https://ui.adsabs.harvard.edu/abs/2013AGUFMED12B..05R/abstract

UNFCCC. (N.D.). Private Sector Initiative actions on adaptation Title of case study Enabling access to weather and climate services in Africa. Retrieved 28 January 2024, from http://business.un.org.

5

Initiatives and Strategies to Empower and Support Farmers

Ankit Kumar Maurya[1], Tripti Bhatia[2] and Raghvendra Singh Gautam[3]

[1&3]Agricultural Economics, Chandra Shekhar Azad University of Agriculture and Technology, Kanpur, Uttar Pradesh
[2]Division of Animal Nutrition, College of Veterinary and Animal Science Bikaner, Rajasthan

---

Abstract

In the contemporary agricultural landscape, empowering farmers stands as a pivotal objective for sustainable development. Various initiatives and strategies have emerged to address the multifaceted challenges faced by farmers worldwide, aiming to enhance their resilience, productivity, and socio-economic well-being. One notable initiative revolves around technological advancement. Integration of innovative technologies such as precision agriculture, IoT-enabled devices, and data analytics empowers farmers by providing real-time insights into crop health, weather patterns, and market trends. Accessible through mobile applications, these tools enable farmers to make informed decisions, optimize resource utilization, and mitigate risks effectively. Additionally, fostering agricultural education and training programs plays a fundamental role in farmer empowerment. By equipping farmers with modern agronomic practices, financial literacy, and business management skills, these programs enhance their capacity to adapt to changing market dynamics and environmental conditions. Furthermore, initiatives promoting sustainable farming practices, including organic farming, conservation agriculture, and agroforestry, not only ensure long-term environmental sustainability but also contribute to the economic viability of smallholder farmers. Collaborative platforms and farmer cooperatives are instrumental in amplifying the voices of farmers and advocating for their rights. By facilitating collective bargaining power, these initiatives enable farmers to negotiate fair prices, access essential inputs at subsidized rates, and engage in value-added activities such as processing and marketing. Moreover, policy interventions that prioritize farmer-centric approaches are indispensable for fostering an enabling environment. Governments and stakeholders need to enact policies that promote land tenure security, provide access to credit facilities, and invest in rural infrastructure development. Therefore, empowering farmers necessitates a multifaceted approach encompassing technological innovation, capacity building, sustainable practices, collective action, and supportive policies. By embracing these initiatives and strategies, the agricultural sector can foster inclusive growth, alleviate poverty, and ensure food security for generations to come.

Keywords: *Empower, Socio-economic wellbeing, Innovative Technology, Policies, etc.*

Introduction

Amidst the vast fields and verdant landscapes lies the heartbeat of nations – the farmers. Yet, despite their indispensable role in feeding the world, farmers often find themselves on the front lines of adversity, grappling with challenges ranging from volatile markets to climate uncertainties. Recognizing the pivotal importance of agriculture in fostering sustainable development, a plethora of initiatives and strategies have emerged to empower and support farmers worldwide. In this era of rapid technological advancement, innovation stands as a beacon of hope for farmers striving to navigate a complex agricultural landscape. From precision agriculture to block chain-enabled supply chains, technological solutions are revolutionizing farming practices, empowering farmers with data-driven insights and decision-making tools. Through real-time monitoring of soil health, weather patterns, and market trends, farmers can optimize resource allocation, minimize risks, and enhance productivity. Moreover, the proliferation of mobile applications tailored for agricultural use has democratized access to critical information, connecting farmers to markets, financial services, and advisory support at their fingertips. Thus, technology serves as a catalyst for transformation, bridging the gap between traditional farming practices and the demands of a modernizing world.

However, empowerment extends beyond technological prowess; it encompasses a holistic approach encompassing education, capacity building, and sustainable practices. Investment in agricultural education and training programs equips farmers with the knowledge and skills needed to adapt to changing conditions and embrace innovative practices. By promoting agro-ecological principles, organic farming, and climate-smart techniques, farmers can enhance resilience, mitigate environmental degradation, and ensure long-term sustainability. Furthermore, initiatives focusing on financial literacy and access to credit empower farmers to invest in inputs, machinery, and infrastructure, unlocking the potential for growth and prosperity.

Sustainable development lies at the heart of farmer empowerment, transcending economic imperatives to encompass social and environmental dimensions. Collective action through farmer cooperatives and community-based organizations amplifies the voices of farmers, fostering solidarity, and enabling them to leverage their collective bargaining power. By pooling resources, sharing knowledge, and engaging in value-added activities such as processing and marketing, farmers can achieve economies of scale and capture a larger share of the value chain. Moreover, advocacy efforts aimed at promoting policies conducive to farmer well-being, such as land tenure security, fair trade practices, and rural infrastructure development, are instrumental in creating an enabling environment for agricultural growth.

Overall, the empowerment of farmers is not merely a moral imperative but a strategic necessity for achieving sustainable development goals. By harnessing the power of innovation, education, collective action, and policy support, stakeholders can ignite a transformative change in the agricultural sector, paving the way for a future where farmers thrive, communities prosper, and food security is assured. As we embark on this journey towards empowerment, let us heed the call to action, igniting the flames of progress and nurturing a brighter tomorrow for generations to come.

Technological Advancement

Technological advancement plays a pivotal role in initiatives and strategies aimed at empowering and supporting farmers. It offers innovative solutions to address various challenges faced by farmers, enhance agricultural productivity, and improve socio-economic outcomes. Here's a discussion on technological advancement within the context of empowering and supporting farmers:

1. **Precision Agriculture:** Precision agriculture involves the use of advanced technologies such as GPS, drones, sensors, and data analytics to optimize farm management practices. These technologies enable farmers to monitor crop health, soil moisture levels, and weather conditions in real-time. By precisely targeting inputs such as water, fertilizers, and pesticides, farmers can minimize wastage, reduce costs, and maximize yields, thereby improving productivity and profitability.

2. **Digital Farming Technologies:** Digital farming technologies encompass a wide range of digital tools and platforms that facilitate farm management tasks, decision-making processes, and communication channels. Mobile applications, farm management software, and online platforms provide farmers with access to agricultural information, market prices, weather forecasts, and advisory services. These technologies enable farmers to

make informed decisions, plan their operations efficiently, and access support services conveniently, regardless of their location.

3. **IoT-enabled Devices:** The Internet of Things (IoT) has revolutionized agriculture by enabling the deployment of IoT-enabled devices such as soil moisture sensors, weather stations, and automated irrigation systems. These devices collect real-time data from the farm environment and transmit it to farmers' smartphones or computers, allowing for timely interventions and optimized resource management. IoT technologies help farmers monitor crop conditions, detect pest infestations, and automate routine tasks, leading to increased efficiency and reduced labor requirements.

4. **Remote Sensing and Satellite Imagery:** Remote sensing technologies, including satellite imagery and aerial drones, provide farmers with valuable insights into crop health, vegetation vigor, and land use patterns. By analyzing satellite images, farmers can identify areas of stress, nutrient deficiencies, or pest outbreaks early on, enabling targeted interventions and preventive measures. Remote sensing technologies also support land mapping, precision farming applications, and environmental monitoring, contributing to sustainable land management practices and resource conservation.

5. **Block chain and Traceability Systems:** Block chain technology offers transparent and secure systems for tracking and tracing agricultural products throughout the supply chain. By implementing block chain-based traceability systems, farmers can enhance product quality, ensure food safety, and gain access to premium markets that value transparency and authenticity. Block chain also facilitates fairer and more transparent transactions, reducing the role of intermediaries and empowering farmers to receive fair prices for their produce.

Overall, technological advancement presents unprecedented opportunities for empowering and supporting farmers in their agricultural endeavors. By leveraging innovative technologies such as precision agriculture, digital farming tools, IoT-enabled devices, remote sensing, and block chain systems, farmers can enhance productivity, optimize resource management, improve market access, and achieve sustainable agricultural development. Therefore, integrating technological advancements into initiatives and strategies for farmer empowerment is essential for unlocking the full potential of agriculture and ensuring the prosperity of farming communities worldwide.

Capacity Building and Education

Capacity building and education are fundamental components of initiatives and strategies aimed at empowering and supporting farmers. These efforts focus on equipping farmers with the knowledge, skills, and resources necessary to adapt to changing agricultural landscapes, improve productivity, and enhance socio-economic well-being. Here's a discussion on capacity building and education within the context of empowering and supporting farmers:

1. **Agricultural Training Programs:** Agricultural training programs provide farmers with opportunities to learn about modern farming techniques, sustainable practices, and innovative technologies. These programs may include workshops, field demonstrations, and hands-on training sessions conducted by agricultural experts, extension agents, or local agricultural organizations. By participating in such programs, farmers gain practical skills and knowledge that enable them to improve crop yields, conserve natural resources, and adopt climate-resilient farming practices.

2. **Financial Literacy and Access to Credit:** Financial literacy programs educate farmers about basic financial management principles, budgeting, saving, and investment strategies. By understanding financial concepts and practices, farmers can make informed decisions about managing their finances, accessing credit, and investing in their farms. Additionally, initiatives aimed at improving farmers' access to credit provide them with the necessary capital to purchase inputs, invest in equipment, and expand their operations, thereby enhancing their productivity and profitability.

3. **Business Management Skills:** Business management training equips farmers with essential skills in marketing, value chain development, entrepreneurship, and farm business planning. By learning how to identify market opportunities, negotiate prices, and add value to their products, farmers can increase their competitiveness and profitability in the marketplace. Business management skills also enable farmers to diversify their income sources, mitigate risks, and build resilient farming enterprises.

4. **ICT and Digital Literacy:** Training programs focused on information and communication technology (ICT) and digital literacy help farmers harness the power of digital tools and platforms for farm management, market access, and knowledge sharing. By learning how to use mobile phones, computers, and internet-based applications, farmers can access

agricultural information, weather forecasts, market prices, and advisory services remotely. ICT training empowers farmers to stay informed, connect with peers and experts, and adapt to the digital age of agriculture.

5. **Gender-Sensitive Training:** Gender-sensitive training programs recognize the unique roles and challenges faced by women farmers and aim to empower them through targeted interventions. These programs provide women farmers with access to education, training, resources, and decision-making opportunities within farming households and communities. By addressing gender disparities and promoting women's participation in agricultural development, gender-sensitive training programs contribute to inclusive and sustainable rural development.

6. **Extension Services:** Agricultural extension services play a crucial role in delivering training, advisory support, and technical assistance to farmers in rural areas. Extension agents, often employed by government agencies or NGOs, work closely with farmers to provide personalized advice, address specific challenges, and facilitate technology adoption. By bridging the gap between research institutions, policymakers, and farmers, extension services contribute to the dissemination of best practices, innovations, and market information, thereby enhancing farmer productivity and resilience.

Sustainable Farming Practices

Sustainable farming practices are integral to initiatives and strategies aimed at empowering and supporting farmers. These practices prioritize environmental stewardship, resource conservation, and resilience to climate change while promoting socio-economic viability for farming communities. Here's a discussion on sustainable farming practices within the context of empowering and supporting farmers:

1. **Organic Farming:** Organic farming practices avoid the use of synthetic pesticides, fertilizers, and genetically modified organisms (GMOs), focusing instead on natural methods of pest control, soil fertility management, and crop rotation. Initiatives promoting organic farming empower farmers to produce healthier, chemical-free food, reduce environmental pollution, and access premium markets that value organic certification.

2. **Agro-ecology:** Agro-ecological practices integrate ecological principles and traditional knowledge into agricultural systems to enhance biodiversity, soil health, and ecosystem resilience. By mimicking natural

ecosystems and promoting biological diversity, agro-ecology supports pest control, nutrient cycling, and soil conservation, reducing farmers' dependence on external inputs and enhancing long-term sustainability.

3. **Conservation Agriculture:** Conservation agriculture practices minimize soil disturbance, promote soil cover, and diversify crop rotations to improve soil structure, moisture retention, and fertility. By reducing erosion, conserving water, and enhancing soil health, conservation agriculture helps farmers adapt to climate change, mitigate environmental degradation, and sustainably increase crop yields over time.

4. **Agroforestry:** Agroforestry integrates trees and shrubs into agricultural landscapes to provide multiple benefits, including shade, windbreaks, soil stabilization, and biodiversity conservation. Agroforestry systems can improve soil fertility, enhance water retention, and diversify farmers' income streams through the production of timber, fruits, nuts, and non-timber forest products.

5. **Integrated Pest Management (IPM):** Integrated pest management combines biological, cultural, and chemical control methods to manage pest populations while minimizing environmental impacts and preserving beneficial insects. IPM strategies include crop diversification, habitat manipulation, biological control agents, and judicious use of pesticides, reducing farmers' reliance on synthetic chemicals and promoting ecological balance.

6. **Water-saving Irrigation Techniques:** Sustainable irrigation practices, such as drip irrigation, sprinkler systems, and rainwater harvesting, help farmers optimize water use, minimize wastage, and cope with water scarcity. By improving water efficiency and reducing irrigation costs, these techniques enhance farmers' resilience to droughts and water stress while conserving precious water resources for future generations.

7. **Crop Diversity and Rotation:** Crop diversity and rotation are essential components of sustainable farming systems, promoting soil health, pest management, and resilience to climate variability. By diversifying crops and rotating them seasonally, farmers can break pest and disease cycles, improve nutrient cycling, and reduce the risk of crop failures due to adverse weather conditions.

Climate Resilience and Adaptation

Climate resilience and adaptation are crucial components of initiatives and strategies aimed at empowering and supporting farmers, particularly in the face of increasing climate variability and extreme weather events. These initiatives focus on equipping farmers with the knowledge, tools, and resources needed to adapt to changing climatic conditions and build resilience in their farming practices. Here's a discussion on climate resilience and adaptation within the context of empowering and supporting farmers:

1. **Climate-smart Agricultural Practices:** Climate-smart agricultural practices integrate climate adaptation, mitigation, and food security objectives to enhance resilience and sustainability in farming systems. These practices include drought-resistant crop varieties, water-saving irrigation techniques, soil conservation measures, and agroforestry systems. By adopting climate-smart practices, farmers can mitigate the impacts of climate change, improve resource use efficiency, and maintain productivity in the face of changing climatic conditions.

2. **Weather Forecasting and Early Warning Systems:** Access to accurate weather forecasts and early warning systems is essential for farmers to anticipate and prepare for weather-related risks such as droughts, floods, and extreme temperatures. Initiatives that provide farmers with timely weather information and advisories enable them to make informed decisions about crop management, irrigation scheduling, and risk mitigation measures, reducing vulnerability to climate-related disasters and crop losses.

3. **Crop Diversification and Resilient Seed Systems:** Crop diversification and the use of resilient seed varieties are key strategies for enhancing climate resilience in agriculture. By planting a diverse range of crops adapted to different climatic conditions and soil types, farmers can spread their risk and buffer against yield losses caused by climate variability. Initiatives that promote resilient seed systems, including the conservation of traditional crop varieties and the development of climate-smart seeds, help farmers adapt to changing climate conditions and maintain food security.

4. **Water Management and Irrigation Infrastructure:** Improved water management and irrigation infrastructure are essential for building climate resilience in agriculture, particularly in regions prone to droughts and water scarcity. Initiatives that promote efficient water use, rainwater harvesting, and small-scale irrigation systems enable farmers

to cope with water stress and ensure crop production during dry periods. Investments in irrigation infrastructure, such as small reservoirs, drip irrigation, and canal rehabilitation, help farmers adapt to changing precipitation patterns and water availability.

5. **Capacity Building and Farmer Training:** Capacity building and farmer training programs play a vital role in enhancing climate resilience and adaptation in agriculture. These programs provide farmers with the knowledge, skills, and tools needed to implement climate-smart practices, manage weather-related risks, and adapt to changing climatic conditions. By empowering farmers with climate-resilient farming techniques and decision-making capabilities, capacity building initiatives contribute to sustainable agriculture and improved livelihoods in the face of climate change.

6. **Insurance and Risk Management:** Climate risk insurance and risk management mechanisms can help farmers cope with climate-related losses and disruptions to their livelihoods. Initiatives that provide access to weather-based insurance products, index-based insurance schemes, and disaster risk financing mechanisms enable farmers to manage climate risks, recover from weather-related disasters, and sustain their agricultural operations in the long term.

Collective Action and Advocacy

Collective action and advocacy play crucial roles in initiatives and strategies aimed at empowering and supporting farmers, particularly in advocating for their rights, addressing common challenges, and leveraging their collective strength to achieve shared goals. Here's a discussion on collective action and advocacy within the context of empowering and supporting farmers:

1. **Farmer Cooperatives and Associations:** Farmer cooperatives and associations are important vehicles for collective action and solidarity among farmers. These organizations enable farmers to pool their resources, share knowledge and information, and collectively negotiate better prices for their produce. By joining forces, farmers can access economies of scale, reduce input costs, and strengthen their bargaining power in the marketplace. Additionally, farmer cooperatives provide platforms for peer support, capacity building, and community development initiatives, empowering farmers to address common challenges and advocate for their interests effectively.

2. **Market Access and Value Chain Development:** Collective action initiatives focused on market access and value chain development enable farmers to capture a larger share of the value chain and increase their incomes. Farmer cooperatives and producer organizations facilitate collective marketing, value-added processing, and direct sales to consumers, bypassing intermediaries and reducing transaction costs. By organizing collectively, farmers can negotiate fair prices, access premium markets, and develop branding and marketing strategies to differentiate their products and enhance their competitiveness.

3. **Policy Advocacy and Lobbying:** Advocacy efforts aimed at influencing agricultural policies and regulations are critical for empowering and supporting farmers. Farmer organizations and advocacy groups play an important role in representing farmers' interests, voicing their concerns, and lobbying policymakers to enact farmer-friendly policies. By advocating for policies that support land tenure security, access to credit, fair trade practices, and rural infrastructure development, farmers can create an enabling environment for agricultural growth and sustainable rural development.

4. **Knowledge Sharing and Peer Learning:** Collective action initiatives facilitate knowledge sharing and peer learning among farmers, enabling them to exchange experiences, best practices, and innovative solutions to common challenges. Farmer field schools, study tours, and farmer-to-farmer extension programs provide opportunities for farmers to learn from each other, adopt new technologies, and improve their farming practices. By sharing knowledge and experiences, farmers can build social capital, strengthen their resilience, and empower themselves to overcome barriers to agricultural development.

5. **Community Development and Social Impact:** Collective action initiatives contribute to community development and social impact by fostering social cohesion, equity, and inclusiveness within farming communities. By working together towards common goals, farmers can address social issues such as gender inequality, youth unemployment, and rural poverty. Collective action initiatives promote participatory decision-making, empower marginalized groups, and build solidarity among farmers, creating opportunities for inclusive growth and sustainable development.

Overall, collective action and advocacy are essential components of initiatives and strategies to empower and support farmers. By organizing collectively, farmers can leverage their collective strength to address common challenges, access markets, influence agricultural policies, and improve their livelihoods. Empowering farmers through collective action and advocacy not only strengthens their position in the agricultural value chain but also contributes to inclusive rural development, social justice, and sustainable agriculture.

Access to Resources

Access to resources is a fundamental aspect of initiatives and strategies aimed at empowering and supporting farmers. These resources include land, water, seeds, fertilizers, credit facilities, and knowledge, among others. Ensuring equitable access to resources enables farmers to improve their productivity, enhance their livelihoods, and achieve sustainable agricultural development. Here's a discussion on access to resources within the context of empowering and supporting farmers:

1. **Land Tenure Security:** Secure land tenure is essential for farmers to invest in their farms, adopt sustainable practices, and plan for the long term. Initiatives aimed at securing land rights for farmers, particularly smallholders and marginalized groups, are crucial for empowering farmers and promoting agricultural development. Land reform programs, land titling initiatives, and community land rights recognition efforts help protect farmers' land rights and ensure equitable access to land resources.
2. **Access to Water:** Access to water is critical for agricultural production, particularly in regions prone to water scarcity and droughts. Initiatives that promote water resource management, irrigation infrastructure development, and water conservation techniques enable farmers to access and utilize water resources more efficiently. By investing in small-scale irrigation systems, rainwater harvesting, and water-saving technologies, farmers can mitigate the impacts of water scarcity and improve crop yields.
3. **Seeds and Planting Materials:** Access to quality seeds and planting materials is essential for farmers to achieve high crop yields and maintain genetic diversity in agricultural systems. Seed distribution programs, seed banks, and community seed saving initiatives enable farmers to access a diverse range of locally adapted crop varieties suited to their agro ecological conditions. By preserving traditional crop varieties and promoting seed sovereignty, farmers can reduce dependence on external seed sources and adapt to changing climate conditions.

4. **Access to Inputs and Technologies:** Access to agricultural inputs such as fertilizers, pesticides, and farm machinery is essential for farmers to improve productivity and reduce post-harvest losses. Input subsidy programs, agricultural extension services, and farmer cooperatives facilitate access to inputs, technologies, and agronomic advice for smallholder farmers. By providing access to subsidized inputs, credit facilities, and training on modern farming techniques, these initiatives empower farmers to adopt sustainable practices and increase their agricultural productivity.
5. **Access to Credit and Financial Services:** Access to credit and financial services is crucial for farmers to invest in their farms, purchase inputs, and withstand economic shocks. Microfinance institutions, agricultural credit cooperatives, and rural banking networks provide farmers with access to credit, savings, insurance, and other financial services tailored to their needs. By accessing affordable credit and financial products, farmers can invest in farm improvements, diversify their income sources, and build resilience to economic and climatic uncertainties.
6. **Access to Knowledge and Extension Services:** Access to agricultural knowledge and extension services is essential for farmers to adopt best practices, improve their skills, and stay informed about new technologies and market opportunities. Extension programs, farmer field schools, and digital agriculture platforms provide farmers with access to training, technical assistance, and market information. By investing in extension services and knowledge-sharing initiatives, stakeholders can empower farmers to make informed decisions, adopt sustainable practices, and enhance their livelihoods.

Market Access and Value Chain Development

Market access and value chain development are essential components of initiatives and strategies aimed at empowering and supporting farmers. These initiatives focus on improving farmers' access to markets, reducing post-harvest losses, and enhancing their participation in value chains to increase their incomes and improve their livelihoods. Here's a discussion on market access and value chain development within the context of empowering and supporting farmers:

1. **Market Information Systems:** Access to timely and accurate market information is crucial for farmers to make informed decisions about what crops to grow, when to sell, and where to sell their produce. Market information systems, including mobile applications, agricultural

extension services, and market intelligence platforms, provide farmers with real-time updates on market prices, demand trends, and consumer preferences. By accessing market information, farmers can optimize their production plans, negotiate better prices, and identify lucrative market opportunities, thereby increasing their profitability and competitiveness in the marketplace.

2. **Market Linkages and Networking:** Facilitating market linkages and networking opportunities for farmers enables them to connect with buyers, suppliers, and other stakeholders along the value chain. Farmer cooperatives, producer organizations, and agribusiness networks provide platforms for farmers to collectively market their produce, pool their resources, and access markets at scale. By organizing collectively, farmers can negotiate better prices, reduce transaction costs, and access premium markets that value traceability, quality, and sustainability.
3. **Value Addition and Processing:** Adding value to agricultural products through processing and value-added activities can significantly increase farmers' incomes and create employment opportunities in rural areas. Initiatives that promote value addition, product diversification, and food processing empower farmers to capture a larger share of the value chain and access higher-value markets. By investing in processing facilities, training programs, and market linkages, stakeholders can support farmers in adding value to their produce, reducing post-harvest losses, and enhancing the marketability of their products.
4. **Quality Standards and Certification:** Meeting quality standards and certification requirements is essential for accessing export markets and premium market segments. Initiatives that support farmers in adopting good agricultural practices, quality assurance systems, and certification standards enable them to comply with market requirements and access higher-value markets. By providing training, technical assistance, and support for certification processes, stakeholders can empower farmers to produce high-quality products that meet market demands and command premium prices.
5. **Contract Farming and Market Agreements:** Contract farming arrangements and market agreements provide farmers with guaranteed markets, stable prices, and access to inputs and technical support from agribusiness partners. By entering into contracts with buyers, processors, or exporters, farmers can reduce market risks, secure incomes, and access markets for their produce. Contract farming initiatives also

facilitate technology transfer, knowledge sharing, and capacity building for farmers, enabling them to adopt best practices and improve their productivity and competitiveness.

6. **Infrastructure Development:** Investing in rural infrastructure such as roads, cold storage facilities, packing houses, and market hubs is essential for improving market access and value chain efficiency. Infrastructure development initiatives reduce transportation costs, improve product quality, and extend the shelf life of perishable produce, thereby increasing farmers' access to distant markets and reducing post-harvest losses. By investing in infrastructure, stakeholders can unlock the economic potential of rural areas, create employment opportunities, and stimulate agricultural growth and development.

Gender and Social Inclusion

Gender and social inclusion are critical considerations in initiatives and strategies aimed at empowering and supporting farmers. These initiatives focus on addressing gender disparities, promoting women's empowerment, and ensuring the inclusion of marginalized groups in agricultural development efforts. Here's a discussion on gender and social inclusion within the context of empowering and supporting farmers:

1. **Women's Empowerment in Agriculture:** Women play a significant role in agriculture, yet they often face systemic barriers that limit their access to resources, decision-making power, and economic opportunities. Initiatives that promote women's empowerment in agriculture aim to address these barriers by providing women farmers with access to land, credit, training, and extension services. By empowering women farmers, these initiatives contribute to improved household nutrition, food security, and poverty reduction, as women typically reinvest a larger portion of their income into their families and communities.

2. **Gender-Responsive Extension Services:** Gender-responsive extension services recognize the unique needs and constraints faced by women farmers and tailor agricultural training and advisory support accordingly. Female extension agents, women-led farmer groups, and gender-sensitive training programs provide women farmers with a safe and inclusive space to access agricultural information, acquire new skills, and participate in decision-making processes. By ensuring the inclusion of women in extension services, these initiatives contribute to more equitable and sustainable agricultural development outcomes.

3. **Access to Land and Property Rights:** Secure land tenure is essential for women's empowerment and agricultural development, yet women often face discriminatory land tenure laws and practices that limit their access to land and property rights. Initiatives that promote gender-equitable land tenure systems, land redistribution programs, and legal reforms empower women to access and control land resources, make productive investments, and participate in decision-making processes related to land management and use.

4. **Financial Inclusion for Women Farmers:** Access to financial services such as credit, savings, and insurance is critical for women farmers to invest in their farms, access inputs, and cope with economic shocks. Gender-sensitive financial inclusion initiatives aim to address the barriers that prevent women from accessing formal financial services, including lack of collateral, limited financial literacy, and discriminatory lending practices. By providing women farmers with access to financial products tailored to their needs, these initiatives enhance women's economic resilience and enable them to participate more effectively in agricultural value chains.

5. **Promotion of Gender-Responsive Policies:** Advocacy efforts aimed at promoting gender-responsive agricultural policies and programs are essential for addressing gender disparities and promoting women's empowerment in agriculture. Gender mainstreaming in agricultural policies, gender budgeting, and the inclusion of gender-specific indicators in monitoring and evaluation frameworks help ensure that agricultural interventions benefit women as well as men. By advocating for gender-equitable policies and programs, stakeholders can create an enabling environment for women's empowerment and social inclusion in agriculture.

6. **Youth Inclusion in Agriculture:** Youth inclusion initiatives aim to engage young people in agriculture by providing them with access to education, training, employment opportunities, and support for entrepreneurship. By promoting youth participation in agriculture, these initiatives contribute to intergenerational knowledge transfer, innovation, and the sustainability of agricultural livelihoods. Additionally, youth inclusion in agriculture helps address the aging farmer population and ensures the continuity of farming traditions and practices.

Rural Infrastructure Development

Rural infrastructure development is a vital aspect of initiatives and strategies aimed at empowering and supporting farmers. It encompasses the construction and improvement of roads, irrigation systems, storage facilities, market infrastructure, and other essential amenities in rural areas. Enhancing rural infrastructure facilitates' access to markets, inputs, and services, improves agricultural productivity, and promotes overall rural development. Here's a discussion on rural infrastructure development within the context of empowering and supporting farmers:

1. **Improved Transportation Networks:** Developing and upgrading rural road networks is essential for connecting farmers to markets, input suppliers, and other essential services. All-weather roads reduce transportation costs, facilitate the timely movement of agricultural produce, and enable farmers to access markets even during adverse weather conditions. Initiatives that focus on road construction, rehabilitation, and maintenance help overcome transportation barriers, improve market access, and increase farmers' competitiveness in the marketplace.

2. **Enhanced Access to Irrigation:** Access to reliable irrigation infrastructure is critical for mitigating the impacts of climate variability, increasing crop yields, and ensuring food security in rural areas. Investing in irrigation infrastructure, such as small-scale irrigation schemes, canal rehabilitation, and water storage facilities, enables farmers to cultivate crops year-round, diversify their production, and increase their incomes. By providing farmers with access to water for irrigation, infrastructure development initiatives contribute to improved agricultural productivity and resilience to climate change.

3. **Storage and Post-Harvest Facilities:** Developing storage and post-harvest facilities in rural areas helps reduce post-harvest losses, improve market access, and enhance farmers' incomes. Cold storage facilities, warehouses, and agro-processing centers enable farmers to store perishable produce, add value to their products, and access markets beyond the harvest season. Infrastructure development initiatives that focus on building storage facilities, improving logistics, and promoting value-added processing contribute to increased farm gate prices and improved livelihoods for farmers.

4. **Market Infrastructure and Facilities:** Enhancing market infrastructure and facilities, such as market sheds, trading centers, and wholesale

markets, creates conducive environments for buying and selling agricultural produce. Well-equipped market infrastructure improves market transparency, reduces transaction costs, and fosters competitive pricing mechanisms. Initiatives that invest in market infrastructure development promote vibrant agricultural markets, facilitate trade, and stimulate economic growth in rural areas.

5. **Access to Financial Services:** Access to financial services such as banking, credit, and insurance is essential for rural development and agricultural investment. Establishing rural banking networks, microfinance institutions, and agricultural credit cooperatives expands farmers' access to credit, savings, and insurance products tailored to their needs. Financial inclusion initiatives that focus on rural infrastructure development enable farmers to invest in farm improvements, purchase inputs, and cope with economic shocks, thereby enhancing their productivity and resilience.

6. **Information and Communication Technology (ICT) Infrastructure:** Investing in ICT infrastructure, such as mobile networks, internet connectivity, and digital platforms, enhances farmers' access to agricultural information, market prices, and advisory services. ICT infrastructure enables farmers to access weather forecasts, market information, and extension services remotely, facilitating informed decision-making and knowledge sharing. Infrastructure development initiatives that promote ICT connectivity in rural areas bridge the digital divide, empower farmers with technology-enabled solutions, and enhance their productivity and resilience.

Youth Engagement in Agriculture

Youth engagement in agriculture is crucial for ensuring the sustainability and viability of the agricultural sector, as well as promoting rural development. Initiatives and strategies that focus on empowering and supporting young people to participate in agriculture can drive innovation, enhance productivity, and address demographic challenges in rural areas. Here's a discussion on youth engagement in agriculture within the context of empowering and supporting farmers:

1. **Training and Capacity Building:** Providing young people with access to agricultural training, technical education, and capacity-building programs equips them with the skills and knowledge needed to succeed in agriculture. Hands-on training, apprenticeships, and vocational courses in agronomy, livestock management, agribusiness, and

sustainable farming practices prepare youth for careers in agriculture and entrepreneurship.

2. **Access to Land and Resources:** Ensuring young farmers have access to land, credit, inputs, and other essential resources is critical for overcoming barriers to entry into agriculture. Land leasing programs, land redistribution schemes, and youth-friendly credit facilities enable young people to start and expand their farming enterprises. Access to resources empowers youth to pursue agricultural ventures, innovate, and contribute to agricultural development.
3. **Support for Agribusiness and Entrepreneurship:** Promoting agribusiness and entrepreneurship among young people encourages innovation, value addition, and market-driven approaches to agriculture. Providing support for agribusiness startups, incubation programs, and entrepreneurship training helps young farmers develop business skills, access markets, and create sustainable agricultural enterprises. Initiatives that facilitate access to finance, market information, and mentorship empower youth to pursue agricultural entrepreneurship as a viable career option.
4. **ICT and Digital Solutions:** Leveraging information and communication technology (ICT) and digital solutions can attract and engage young people in agriculture. Digital platforms, mobile applications, and e-commerce platforms provide youth with access to agricultural information, market opportunities, and extension services. ICT-enabled solutions enhance youth participation in agriculture by facilitating access to knowledge, networking, and market linkages.
5. **Youth-Led Organizations and Networks:** Supporting youth-led organizations, networks, and movements strengthens youth participation and advocacy in agriculture. Youth farmer associations, cooperatives, and advocacy groups provide platforms for young farmers to voice their concerns, share experiences, and influence agricultural policies. Engaging youth in decision-making processes and leadership roles empowers them to drive positive change and contribute to agricultural development agendas.
6. **Promotion of Sustainable and Climate-Smart Practices:** Encouraging youth involvement in sustainable and climate-smart agriculture promotes environmental stewardship and resilience to climate change. Training programs, demonstration farms, and peer-to-peer learning initiatives educate young farmers about sustainable farming practices,

conservation agriculture, and climate adaptation strategies. Empowering youth to adopt sustainable practices ensures the long-term viability of agriculture and fosters environmental sustainability.

Policy Support and Enabling Environment

Policy support and creating an enabling environment are crucial elements of initiatives and strategies aimed at empowering and supporting farmers. Policies that prioritize agriculture, provide necessary resources, and create conducive regulatory frameworks play a vital role in enhancing farmers' productivity, livelihoods, and overall well-being. Here's a discussion on policy support and enabling environment within the context of empowering and supporting farmers:

1. **Agricultural Investment:** Governments can promote agricultural development by allocating sufficient budgetary resources to the sector, investing in rural infrastructure, research and development, and extension services. Policy support for agricultural investment enables farmers to access essential resources, technologies, and services needed to increase productivity, enhance market access, and improve their livelihoods.
2. **Market Access and Trade Policies:** Policies that facilitate market access, reduce trade barriers, and promote fair trade practices are essential for empowering farmers and enhancing their competitiveness in domestic and international markets. Governments can support farmers by negotiating favorable trade agreements, providing market information, and implementing measures to ensure fair prices and market transparency.
3. **Land Tenure Security:** Secure land tenure rights are critical for farmers to invest in their land, adopt long-term planning, and access credit and other financial services. Policies that promote land tenure security, land redistribution, and land reform initiatives empower farmers to access and control land resources, make productive investments, and improve their livelihoods.
4. **Access to Finance and Credit:** Policies that promote financial inclusion, provide access to credit, and offer risk management tools are essential for supporting farmers' investment and entrepreneurial activities. Governments can support farmers by establishing agricultural credit facilities, microfinance programs, and insurance schemes tailored to their needs. Access to finance enables farmers to invest in inputs, technology, and infrastructure, thereby increasing productivity and resilience.

5. **Technology and Innovation Policies:** Policies that promote agricultural research and innovation, technology transfer, and adoption of sustainable farming practices empower farmers to increase productivity, reduce costs, and adapt to changing environmental conditions. Governments can support farmers by investing in agricultural research institutions, promoting technology adoption, and providing incentives for sustainable practices.
6. **Social Protection and Safety Nets:** Social protection programs, safety nets, and insurance schemes provide a safety net for farmers during times of hardship, such as natural disasters, market fluctuations, or economic shocks. Policies that promote social protection empower farmers to cope with risks, reduce vulnerability, and maintain their livelihoods in the face of adversity.

Conclusion

In conclusion, initiatives and strategies to empower and support farmers are essential for promoting sustainable agricultural development, enhancing food security, and improving rural livelihoods. By prioritizing investments in rural infrastructure, access to resources, market access and capacity building, and policy support, stakeholders can create enabling environments that empower farmers to thrive. Embracing sustainable farming practices, promoting gender and social inclusion, and engaging youth in agriculture are key pillars of empowerment strategies that ensure the resilience and prosperity of farming communities. Moreover, recognizing the diverse needs and challenges faced by farmers, including smallholders, women, and marginalized groups, is crucial for designing inclusive and effective interventions. By fostering partnerships, collaboration, and innovation, stakeholders can amplify the impact of empowerment initiatives and promote holistic approaches to agricultural development. Ultimately, empowering farmers not only strengthens the agricultural sector but also contributes to broader socio-economic development, environmental sustainability and inclusive growth in rural areas. It is imperative that we continue to prioritize the empowerment and support of farmers as they are the backbone of our food systems and stewards of our land.

References

Angom, J. and Viswanathan, P.K. 2023. Climate-Smart Agricultural Practices and Technologies in India and South Africa: Implications for Climate Change Adaption and Sustainable Livelihoods. In The Palgrave Handbook of Socio-ecological Resilience in the Face of Climate Change: Contexts from a Developing Country (pp. 161-195). Singapore: Springer Nature Singapore.

Bihari, B., Singh, M., Bishnoi, R. and Mishra, P.K. 2019. Issues, challenges and strategies for doubling the farmers' income in India–A review. The Indian Journal of Agricultural Sciences, 89(8): 1219-1224.

Gereffi, G. 2019. Global value chains, development, and emerging economies 1. In Business and Development Studies: 125-158.

Goldar, A., Sharma, S., Sawant, V. and Jain, S. 2019. Climate change & technology transfer: Barriers, technologies and mechanisms (No. 382). Working Paper.

Mir, D.A., Doll, C.N., Lindner, R. and Parray, M.T. 2020. Explaining the diffusion of energy-efficient lighting in India: A technology innovation systems approach. Energies, 13(21): 5821.

Pandey, V., Nagarajan, H.K. and Kumar, D. 2021. Impact of Gendered Participation in market-linked value-chains on Economic Outcomes: Evidence from India. Food Policy, 104: 102142.

Shrestha, R.B. and Thapa, Y.B. 2019. Policy and Program Priorities in Agricultural Research & Development in South Asia. Agricultural Policy and Program Framework: Priority Areas for Research & Development in South Asia, p.1.

Tandon, I. 2019. Gendering Farmer Producer Companies at the Agricultural Frontier of India: Empowerment or Burden?. Commodity Frontiers and Global Capitalist Expansion: Social, Ecological and Political Implications from the Nineteenth Century to the Present Day, 79-109.

Ton, G., de Grip, K., Lançon, F., Onumah, G.E. and Proctor, F.J. 2014. Empowering smallholder farmers in markets: Strengthening the advocacy capacities of national farmer organisations through collaborative research. Food Security, 6: 261-273.

Wani, S.P., Bergvinson, D., Raju, K.V., Gaur, P.M. and Varshney, R.K. 2016. Mission India for transforming agriculture (MITrA).

[illegible] Singh, A., [illegible] Kumar, [illegible] 2017. [illegible] challenges and strategies for doubling the farmers' income in India: A review. The Indian Journal of Agricultural Sciences, 87 (11), [illegible].
[illegible] 2019. [illegible] development and emerging economies. [illegible] and Development Studies, [illegible].
[illegible] 2018. Climate change and technology transfer [illegible] Working Paper.
[illegible] 2020. [illegible] diffusion of energy [illegible] innovation [illegible] 1321.
[illegible] Value [illegible] 2019. [illegible] linked value chain [illegible] India. Food Policy, 101 [illegible].
[illegible] and [illegible] and Program Priorities [illegible] Developments in South Asia [illegible] Policy and Program Framework: Priority Areas for Research & Development [illegible].
[illegible] Agricultural [illegible] of India [illegible] Contemporary [illegible] and [illegible] Social, Economic and Political [illegible] Historical Context to the Present Day, [illegible].
[illegible] farmers' [illegible] Food Security, 6, [illegible].
Wani, S.P., [illegible] 2016. [illegible] transforming agriculture [illegible].

6

Market Access and Supply Chain Innovation for Agricultural Produce

***Ritesh Singh Gangwar*[1]*, Ankit Kumar Maurya*[2]**

[1]Subject Matter Specialist (Agronomy), KVK, Chandauli, Uttar Pradesh
[2]Agricultural Economics, Chandra Shekhar Azad University of Agriculture and Technology, Kanpur, Uttar Pradesh

Abstract

Market access and supply chain innovation play a pivotal role in ensuring the success and sustainability of agricultural produce. In today's globalized economy, market access refers to the ability of farmers to reach consumers both domestically and internationally. It encompasses various factors such as trade agreements, tariffs, regulations, and infrastructure. Access to diverse markets not only expands the customer base but also provides opportunities for farmers to fetch better prices for their products. Supply chain innovation complements market access by enhancing efficiency and reliability throughout the agricultural value chain. This innovation involves the adoption of advanced technologies, streamlined logistics, and sustainable practices to optimize production, storage, transportation, and distribution processes. By leveraging innovations such as IoT sensors, block chain and data analytics, farmers can better monitor crops, reduce waste, and ensure product quality. Additionally, advancements in cold chain logistics enable perishable goods to reach distant markets while maintaining freshness. The importance of market access and supply chain innovation in agriculture cannot be overstated. They enable farmers to respond to changing consumer demands, mitigate risks associated with climate change and market fluctuations, and improve overall profitability. Moreover, by connecting farmers directly with consumers through e-commerce platforms and farmer's markets, market access and supply chain innovation empower small-scale producers to compete in the global marketplace. Therefore, market access and supply chain innovation are imperative for the sustainable growth of the agricultural sector. Governments, industry stakeholders, and farmers must collaborate to remove barriers to market entry, invest in infrastructure, and foster a culture of innovation to ensure a resilient and prosperous agricultural industry that can meet the demands of a growing population while safeguarding environmental and social welfare.

Keywords: *Agricultural Produce, Market, Innovation, Transportation, Distribution, etc.*

Introduction

Market access and supply chain innovation are two critical pillars that underpin the success and sustainability of agricultural produce in today's interconnected world. As the global population continues to grow, along with increasing consumer demands and evolving market dynamics, ensuring efficient access to markets and implementing innovative supply chain practices have become burning imperatives for agricultural producers worldwide.

Market access refers to the ability of farmers and agricultural producers to effectively reach consumers, both domestically and internationally. It encompasses a multitude of factors, including trade agreements, tariffs, regulatory frameworks, transportation infrastructure, and market information accessibility. Access to diverse markets not only expands the customer base for agricultural products but also provides opportunities for farmers to capture better prices and maximize their returns on investment. However, challenges such as trade barriers, tariffs, sanitary and phytosanitary regulations, and transportation constraints can hinder market access, limiting the potential for growth and profitability within the agricultural sector.

Supply chain innovation complements market access by revolutionizing the processes involved in the production, storage, transportation, and distribution of agricultural goods. In recent years, advancements in technology, logistics, and sustainability practices have fueled significant innovation within agricultural supply chains. From precision agriculture techniques that optimize resource utilization to the adoption of block chain technology for transparent and traceable supply chains, innovations are driving efficiency, reliability, and resilience throughout the agricultural value chain.

One of the key drivers of supply chain innovation in agriculture is the adoption of digital technologies and data-driven solutions. Internet of Things (IoT) sensors, drones, satellite imagery, and other digital tools provide farmers with real-time insights into crop health, soil conditions, and weather patterns, enabling them to make informed decisions and optimize their agricultural practices. Furthermore, block chain technology is revolutionizing supply chain transparency and traceability by securely recording transactions and product movements from farm to fork, thereby enhancing food safety and consumer trust. Moreover, supply chain innovation encompasses advancements in logistics and transportation, such as cold chain infrastructure and last-mile delivery solutions. These innovations are particularly crucial for perishable agricultural products, ensuring that they reach consumers in optimal condition while minimizing waste and spoilage. Additionally, sustainable supply chain practices, including organic farming methods, renewable energy sources, and eco-friendly packaging, are gaining traction as consumers increasingly prioritize environmental stewardship and ethical sourcing.

Therefore, market access and supply chain innovation are indispensable drivers of growth, competitiveness, and sustainability in the agricultural sector. By facilitating access to diverse markets and implementing cutting-edge supply chain practices, agricultural producers can unlock new opportunities, mitigate risks, and enhance the efficiency and resilience of their operations. Governments, industry stakeholders, and agricultural producers must collaborate to address barriers to market entry, invest in infrastructure and technology, and promote a culture of innovation to ensure the long-term viability of the agricultural sector in meeting the demands of a rapidly changing world.

Trade Policies and Agreements

Trade policies and agreements play a pivotal role in shaping the global agricultural landscape, influencing market access and driving supply chain innovation. In an interconnected world, where agricultural produce traverses international borders, the regulatory frameworks established through these agreements have significant implications for farmers, traders, and consumers alike.

1. **Market Access Facilitation:** Trade policies and agreements facilitate market access by reducing tariffs, quotas, and non-tariff barriers that impede the flow of agricultural goods. Through preferential trade agreements, such as free trade agreements (FTAs) or regional trade blocs, countries negotiate mutually beneficial terms to expand market opportunities for agricultural exporters. By enhancing market access, these agreements stimulate agricultural trade, promote economic growth, and increase farmers' income.

2. **Supply Chain Innovation:** Trade policies also incentivize supply chain innovation by promoting investment in infrastructure, transportation, and logistics. Agreements often include provisions for trade facilitation measures, such as streamlined customs procedures and harmonized regulations, which lower transaction costs and improve the efficiency

of agricultural supply chains. Additionally, agreements may encourage the adoption of technology and best practices in storage, packaging, and distribution, enhancing the quality and safety of agricultural produce.

3. **Regulatory Harmonization:** Harmonizing regulatory standards is a critical aspect of trade agreements, particularly in the agricultural sector where food safety and quality standards vary across jurisdictions. Mutual recognition agreements and regulatory coherence mechanisms ensure that agricultural products meet common standards, facilitating cross-border trade and reducing compliance burdens for exporters. By harmonizing regulations, trade policies promote transparency, predictability, and trust in agricultural markets.

5. **Sustainable Trade Practices:** Increasingly, trade agreements incorporate provisions for sustainable trade practices, promoting environmental stewardship, social responsibility, and fair labor practices. Agreements may include clauses on environmental protection, biodiversity conservation, and climate change mitigation, encouraging responsible production and consumption of agricultural goods. Furthermore, trade policies may support the integration of smallholder farmers into global value chains, ensuring inclusive and equitable benefits from international trade.

Overall, trade policies and agreements play a vital role in facilitating market access and driving supply chain innovation for agricultural produce. By promoting trade liberalization, regulatory harmonization, and sustainable practices, these agreements contribute to the resilience, competitiveness, and sustainability of agricultural sectors worldwide. However, policymakers must balance trade objectives with social and environmental considerations to ensure that trade benefits are equitably distributed and contribute to the long-term well-being of farming communities and the planet.

Regulatory Environment

The regulatory environment plays a critical role in determining market access and shaping supply chains for agricultural produce. Regulations governing food safety, quality standards, labeling requirements, and trade barriers directly impact the ability of agricultural producers to access markets and ensure the smooth flow of goods along the supply chain. Understanding and complying with these regulations are essential for agricultural producers to navigate the complexities of domestic and international trade. Food safety regulations are among the most stringent and widely enforced regulations in the agricultural sector. Governments worldwide implement regulations to protect consumers from foodborne illnesses and ensure the safety of agricultural products. Compliance with food safety standards, such as Hazard Analysis and Critical Control Points (HACCP) and Good Agricultural Practices (GAP), is often a prerequisite for market access. Failure to meet these standards can result in product recalls, trade restrictions, and reputational damage, severely impacting market access and consumer trust.

Quality standards and grading systems are another aspect of the regulatory environment that affects market access and supply chains for agricultural produce. Standards for product quality, size, color, and freshness vary between countries and regions, influencing trade negotiations and market preferences. Agricultural producers must adhere to these standards to meet buyer requirements and access premium markets with higher quality standards. Labeling requirements are crucial for market access and consumer transparency. Regulations governing product labeling, including country of origin labeling (COOL), organic certification, and allergen labeling, provide consumers with essential information about the products they purchase. Compliance with labeling requirements is essential for agricultural producers to meet regulatory obligations and gain consumer trust, particularly in markets where transparency and traceability are valued. Trade barriers, including tariffs, quotas, and import/export regulations, also shape the regulatory environment for agricultural trade. Governments use trade barriers to protect domestic industries, ensure food security, and address environmental and social concerns. Negotiating favorable trade agreements and reducing trade barriers is essential for agricultural producers to access foreign markets and compete effectively on a global scale.

In addition to regulatory compliance, supply chain resilience and agility are crucial for navigating regulatory challenges and ensuring uninterrupted flow of agricultural products to markets. Proactive risk management strategies, such as diversifying supply chain sources, investing in infrastructure and technology, and fostering collaborative relationships with suppliers and regulators, can help agricultural producers mitigate regulatory risks and enhance supply chain resilience. The regulatory environment significantly influences market access and supply chain dynamics for agricultural produce. Food safety regulations, quality standards, labeling requirements, and trade barriers shape the regulatory landscape, impacting the ability of agricultural producers to access markets and comply with regulatory obligations. Navigating regulatory complexities and ensuring compliance are essential for agricultural producers to succeed in today's competitive and highly regulated global marketplace.

Infrastructure Development

Infrastructure development plays a crucial role in facilitating market access and driving supply chain innovation for agricultural produce. Robust infrastructure networks spanning transportation, storage, and logistics are essential for connecting farmers to markets, reducing post-harvest losses, and improving the efficiency of agricultural supply chains.

1. **Transportation Networks:** Efficient transportation infrastructure, including roads, railways, ports, and airports, is essential for connecting agricultural production areas to domestic and international markets. Well-maintained roads and highways enable farmers to transport their produce to distribution centers and processing facilities in a timely manner. Additionally, efficient rail and maritime transport systems facilitate the movement of bulk agricultural commodities over long distances, reducing transportation costs and improving market access for farmers located in landlocked regions or remote areas.

2. **Storage and Warehousing Facilities:** Adequate storage and warehousing facilities are essential for maintaining the quality and freshness of agricultural produce throughout the supply chain. Cold storage facilities, silos, and warehouses help prevent post-harvest losses by providing a controlled environment for perishable goods. Infrastructure investments in modern storage technologies, such as temperature-controlled warehouses and refrigerated transport vehicles, enable farmers to extend the shelf life of their products and access distant markets with higher quality produce.

3. **Market Infrastructure:** Market infrastructure, including wholesale markets, terminal markets, and Agri-food parks, serves as vital nodes in the agricultural supply chain, facilitating the aggregation, processing, and distribution of agricultural produce. Investments in modern market infrastructure help create efficient trading platforms where farmers can connect with buyers, negotiate prices, and access value-added services such as sorting, grading, and packaging. Additionally, the development of e-commerce platforms and digital marketplaces enables farmers to access markets beyond geographical boundaries, fostering trade and innovation in agricultural supply chains.

4. **Technology and Innovation Hubs:** Infrastructure investments in technology and innovation hubs support the adoption of digital solutions and advanced agricultural practices across the supply chain. These hubs serve as centers of research, training, and knowledge exchange,

where farmers can access information on best practices, market trends, and technological innovations. By fostering collaboration between researchers, entrepreneurs, and policymakers, technology and innovation hubs drive supply chain innovation, improve productivity, and enhance the competitiveness of agricultural sectors.

5. **Rural Connectivity and Access to Information:** Investments in rural connectivity, including telecommunications infrastructure and internet connectivity, are essential for bridging the digital divide and providing farmers with access to real-time market information, weather forecasts, and agricultural advisory services. By leveraging mobile technology and data analytics, farmers can make informed decisions regarding crop selection, pricing, and marketing strategies, thereby enhancing their market access and competitiveness.

Therefore, infrastructure development is critical for enhancing market access and driving supply chain innovation in agricultural produce. By investing in transportation networks, storage facilities, market infrastructure, technology hubs, and rural connectivity, policymakers can create an enabling environment that empowers farmers, reduces post-harvest losses, and promotes inclusive growth in agricultural sectors. Furthermore, coordinated efforts between governments, private sector actors, and development partners are essential for ensuring sustainable and resilient agricultural supply chains that can meet the evolving needs of global markets.

Market Information Systems

In the realm of agricultural production and distribution, have emerged as powerful tools driving efficiency, transparency, and accessibility. These systems integrate various data sources and analytical tools to provide stakeholders with real-time information, enabling informed decision-making and facilitating market access and supply chain innovation.

1. **Enhancing Market Access:** MIS plays a pivotal role in improving market access for agricultural producers by bridging information gaps between farmers and buyers. Through comprehensive data on market prices, demand trends, and consumer preferences, farmers can make informed decisions on what to produce, when to sell, and where to sell their produce. This ensures optimal pricing and reduces the risk of market volatility, thereby empowering farmers to access broader markets beyond local boundaries.

2. **Facilitating Supply Chain Innovation:** In the realm of supply chain management, MIS facilitates innovation by providing insights into logistical efficiencies, inventory management, and demand forecasting. By leveraging data analytics and predictive modeling, supply chain actors can optimize routes, reduce transportation costs, minimize wastage, and ensure timely delivery of agricultural produce to end consumers. Moreover, MIS enables the integration of emerging technologies such as IoT sensors, block chain, and AI-driven predictive analytics, fostering greater transparency and traceability throughout the supply chain.
3. **Empowering Stakeholders:** MIS empowers various stakeholders across the agricultural value chain, including farmers, traders, processors, and consumers. Farmers gain access to market information that enables them to negotiate fair prices and identify lucrative market opportunities. Traders and processors can optimize procurement strategies and production planning based on real-time demand signals. Additionally, consumers benefit from improved product quality, safety, and traceability, fostering trust and loyalty towards agricultural products.
4. **Challenges and Opportunities:** Despite its transformative potential, the adoption of MIS faces challenges such as limited access to technology, inadequate infrastructure, and data privacy concerns. Addressing these challenges requires collaborative efforts from governments, private sector entities, and development organizations to invest in digital infrastructure, promote data literacy, and establish regulatory frameworks for data sharing and privacy protection. Moreover, there are opportunities to further enhance MIS through advancements in data analytics, cloud computing, and mobile technologies, thereby unlocking new avenues for market access and supply chain innovation in the agricultural sector.

Digital Technologies

Digital technologies have emerged as powerful tools in revolutionizing market access and supply chain innovation for agricultural produce. By leveraging data analytics, mobile applications, block chain, and internet-enabled platforms, digital technologies are transforming traditional agricultural supply chains, enhancing efficiency, transparency, and resilience.

1. **E-commerce and Online Marketplaces:** E-commerce platforms and online marketplaces connect farmers directly with consumers, bypassing traditional intermediaries and reducing transaction costs. Through these platforms, farmers can showcase their products, reach a

broader customer base, and receive payments securely. E-commerce also facilitates traceability and transparency in the supply chain, allowing consumers to trace the origin of agricultural products and verify their authenticity.

2. **Supply Chain Traceability and Quality Assurance:** Block chain technology enables supply chain traceability by recording transactions in a transparent and immutable ledger. By tracking the movement of agricultural produce from farm to fork, block chain enhances food safety, quality assurance, and compliance with regulatory standards. Consumers can verify the authenticity and integrity of agricultural products, while farmers benefit from improved market access and premium prices for certified products.
3. **Precision Agriculture and IoT Solutions:** Digital technologies such as precision agriculture and Internet of Things (IoT) solutions optimize resource management and improve productivity on farms. IoT sensors collect real-time data on soil moisture, weather conditions, and crop health, enabling farmers to make data-driven decisions regarding irrigation, fertilization, and pest management. Precision agriculture enhances crop yields, reduces input costs, and minimizes environmental impact, contributing to sustainable agricultural practices.
4. **Digital Payment and Financial Inclusion:** Digital payment solutions facilitate financial inclusion for farmers by providing access to formal banking services and credit facilities. Mobile payment platforms enable farmers to receive payments for their produce directly into their bank accounts, eliminating the need for cash transactions and reducing the risk of theft or fraud. Digital finance also facilitates access to microloans, insurance products, and savings accounts, empowering farmers to invest in their businesses and improve their livelihoods.

In conclusion, digital technologies are driving transformative changes in market access and supply chain innovation for agricultural produce. By harnessing the power of data, connectivity, and automation, these technologies empower farmers, enhance transparency, and create new opportunities for value creation in agricultural supply chains. Policymakers, private sector actors, and development partners must collaborate to promote the adoption of digital solutions and ensure their equitable benefits reach smallholder farmers and marginalized communities.

Logistics and Distribution

Logistics and distribution play a crucial role in enabling market access and driving supply chain innovation for agricultural produce. Efficient transportation, storage, and distribution networks are essential for connecting producers to consumers, optimizing supply chain operations, and ensuring the timely delivery of fresh and high-quality agricultural products to markets worldwide. One of the primary challenges in agricultural logistics is the perishable nature of many agricultural products. Fruits, vegetables, dairy products, and other perishable goods require careful handling and transportation to maintain their freshness and quality. Cold chain logistics, which involve the transportation and storage of products under controlled temperature conditions, are essential for preserving the shelf life of perishable agricultural produce. Investing in refrigerated trucks, cold storage facilities, and temperature monitoring systems is critical for minimizing post-harvest losses and ensuring that agricultural products reach consumers in optimal condition.

Moreover, last-mile delivery solutions are crucial for connecting agricultural producers to consumers, particularly in urban areas. Efficient last-mile delivery networks, including e-commerce platforms, online marketplaces, and direct-to-consumer delivery services, enable consumers to access a wide range of agricultural products conveniently. By leveraging technology and data analytics, last-mile delivery solutions optimize route planning, reduce delivery times, and enhance customer satisfaction, thereby improving market access and competitiveness for agricultural producers.

In addition to transportation, storage facilities play a vital role in agricultural logistics. Adequate storage capacity, including warehouses, silos, and cold storage facilities, is essential for managing inventory, buffering supply fluctuations, and ensuring product quality throughout the supply chain. Investing in modern storage facilities equipped with temperature and humidity control systems, inventory management software, and automated handling equipment improves supply chain efficiency, reduces waste, and enhances market access for agricultural producers. Furthermore, distribution networks must be flexible and responsive to changing market conditions and consumer preferences. Just-in-time (JIT) delivery systems, which involve delivering products to market precisely when needed, minimize inventory holding costs and reduce supply chain lead times, enhancing market responsiveness and competitiveness. Additionally, collaborative distribution models, such as shared transportation and warehousing arrangements, enable multiple stakeholders to optimize resource utilization, reduce costs, and improve market access for agricultural producers.

Overall, logistics and distribution are integral components of market access and supply chain innovation for agricultural produce. Efficient transportation, storage, and distribution networks enable agricultural producers to reach diverse markets, optimize supply chain operations, and deliver high-quality products to consumers worldwide. By investing in cold chain logistics, last-mile delivery solutions, modern storage facilities, and collaborative distribution networks, agricultural producers can enhance market access, improve competitiveness, and capture value in an increasingly globalized and interconnected marketplace.

Sustainable Practices

Sustainable practices play a pivotal role in enhancing market access and driving supply chain innovation for agricultural produce. As consumers and stakeholders increasingly prioritize environmental stewardship, social responsibility, and ethical sourcing, adopting sustainable practices throughout the agricultural value chain has become essential for maintaining market competitiveness, ensuring long-term viability, and accessing premium markets. One aspect of sustainable practices in agriculture is the adoption of environmentally friendly farming methods. Organic farming, agro ecology, and regenerative agriculture practices minimize the use of synthetic inputs, promote soil health, and enhance biodiversity. By adopting sustainable farming practices, agricultural producers can reduce their environmental footprint, mitigate climate change impacts, and improve soil fertility and resilience. Furthermore, organic and sustainably produced agricultural products often command premium prices in the market, providing financial incentives for producers to embrace sustainable practices. Moreover, sustainable practices in agriculture extend beyond the farm gate to encompass supply chain sustainability. Sustainable sourcing, fair trade initiatives, and ethical labor practices are essential for ensuring social responsibility and equitable distribution of benefits throughout the supply chain. By partnering with suppliers and stakeholders committed to sustainability, agricultural producers can enhance their market access, build brand reputation, and meet the growing demand for ethically sourced products among consumers.

Technologies such as block chain, IoT, and data analytics enable transparent and traceable supply chains, allowing consumers to verify the origin, quality, and sustainability of agricultural products. Block chain-enabled traceability systems provide real-time visibility into the movement of products from farm to fork, facilitating product authentication, food safety compliance, and sustainability certification. By adopting transparent and traceable supply chain solutions, agricultural producers can build consumer trust, differentiate their products in the market, and access premium markets that value sustainability and transparency. Furthermore, sustainable packaging and waste reduction initiatives are integral to supply chain sustainability in agriculture. Adopting eco-friendly packaging materials, reducing packaging waste, and implementing recycling and circular economy practices minimize environmental impact and enhance resource efficiency throughout the supply chain. Sustainable packaging solutions not only reduce carbon footprint and plastic pollution but also appeal to environmentally conscious consumers, thereby improving market access and brand reputation for agricultural producers.

Overall, sustainable practices are essential for enhancing market access and driving supply chain innovation in the agricultural sector. By embracing environmentally friendly farming methods, promoting social responsibility, and adopting transparent and traceable supply chain solutions, agricultural producers can meet the growing demand for sustainable products, access premium markets, and build resilient and competitive supply chains. Investing in sustainability not only ensures the long-term viability of the agricultural sector but also fosters environmental and social well-being for future generations.

Market Diversification

Market diversification is a strategic approach that aims to expand the reach of agricultural produce into new markets, both domestically and internationally. In the context of market access and supply chain innovation, market diversification plays a crucial role in enhancing resilience, mitigating risks, and unlocking new opportunities for agricultural producers.

1. **Expanding Market Access:** Market diversification broadens the scope of potential markets available to agricultural producers, reducing dependency on specific regions or buyer segments. By exploring diverse market channels such as direct-to-consumer sales, exports, and niche markets, farmers can mitigate the impact of market fluctuations and seasonality, ensuring a more stable income stream. Moreover, market diversification enables producers to cater to varied consumer preferences, dietary trends, and cultural preferences, thereby enhancing market responsiveness and competitiveness.

2. **Driving Supply Chain Innovation:** Market diversification necessitates innovative approaches to supply chain management to accommodate the complexities of serving diverse markets. This includes optimizing transportation routes, adapting packaging requirements, and complying with regulatory standards across different markets. Supply chain

innovation is essential to ensure the efficient movement of agricultural produce from farm to fork, minimizing logistical costs, and maximizing product freshness and quality. Leveraging technology such as block chain, IoT sensors, and data analytics further enhances supply chain visibility, traceability, and resilience.

3. **Exploring New Opportunities:** Market diversification opens doors to new opportunities for agricultural producers, including premium pricing, value-added products, and strategic partnerships. By tapping into niche markets such as organic, fair-trade, or specialty products, farmers can capture higher margins and differentiate themselves from competitors. Moreover, international market diversification enables producers to capitalize on export opportunities, tapping into growing demand in emerging economies or targeting specific market segments with unique product offerings.
4. **Challenges and Strategies:** Market diversification presents challenges such as market entry barriers, trade barriers, and cultural differences, which require strategic planning and collaboration among stakeholders. Developing market intelligence, conducting market research, and building strategic partnerships are essential strategies to navigate these challenges successfully. Moreover, investing in product differentiation, quality assurance, and brand development enhances the competitiveness of agricultural produce in diverse markets, fostering long-term sustainability and growth.

Overall, market diversification is a vital strategy for expanding market access and driving supply chain innovation in the agricultural sector. By embracing diversity in markets, products, and partnerships, agricultural producers can unlock new avenues for growth, resilience, and competitiveness in an increasingly dynamic and interconnected global marketplace.

e-NAM

The e-National Agriculture Market (e-NAM) is a digital platform launched by the Government of India with the aim of revolutionizing agricultural marketing and enhancing market access for farmers. e-NAM facilitates online trading of agricultural commodities across designated market yards (mandis) in India, enabling farmers to access a wider pool of buyers, transparent price discovery, and fair market competition. This innovative platform not only improves market access but also drives supply chain innovation by leveraging technology to streamline agricultural marketing processes and optimize supply chain efficiency.

One of the key benefits of e-NAM is the expansion of market access for farmers. Traditionally, farmers were limited to selling their produce at local mandis, where they often faced price manipulation, collusion, and limited market information. With e-NAM, farmers can register online, upload details of their produce, and participate in transparent, competitive bidding processes across multiple mandis. This enables farmers to access a larger pool of buyers, including traders, processors, and exporters, thereby improving price realization and market access for agricultural produce. Moreover, e-NAM promotes supply chain innovation by digitizing and standardizing agricultural marketing processes. Through the e-NAM platform, all transactions, including bidding, payment, and delivery, are conducted electronically, reducing paperwork, transaction costs, and processing times. Additionally, e-NAM provides real-time market information on prices, arrivals, and demand trends, enabling farmers to make informed decisions and optimize their marketing strategies. By leveraging technology and data analytics, e-NAM enhances supply chain transparency, efficiency, and resilience, driving innovation and competitiveness in agricultural markets.

Furthermore, e-NAM fosters collaboration and integration within agricultural supply chains. The platform facilitates seamless connectivity between farmers, traders, commission agents, market functionaries, and government agencies, enabling efficient information sharing, coordination, and market linkages. This collaborative approach promotes trust, transparency, and accountability within the supply chain, leading to improved market access, better price discovery, and reduced market inefficiencies. Additionally, e-NAM encourages the adoption of best practices, quality standards, and value-added services, further enhancing supply chain innovation and competitiveness. However, despite its potential benefits, e-NAM also faces implementation challenges, including limited internet connectivity, infrastructure constraints, and resistance from traditional market intermediaries. Addressing these challenges requires concerted efforts from governments, policymakers, and stakeholders to invest in digital infrastructure, capacity building, and regulatory reforms. Moreover, ensuring inclusive participation of smallholder farmers, women farmers, and marginalized communities is essential for realizing the full potential of e-NAM in transforming agricultural markets and supply chains.

Risk Management

Risk management is essential in ensuring market access and facilitating supply chain innovation for agricultural produce. Agricultural producers face a myriad of risks, including market volatility, natural disasters, supply chain disruptions, and regulatory uncertainties, which can impact their ability to access markets and sustainably manage their operations. One key aspect of risk management in agricultural market access is diversification. By diversifying their market channels, both domestically and internationally, producers can mitigate the impact of market fluctuations and reduce reliance on specific markets. Additionally, diversification allows producers to access different consumer demographics, adapt to changing consumer preferences, and capitalize on emerging market trends, thereby enhancing their resilience to market risks.

Supply chain innovation also plays a crucial role in risk management for agricultural produce. Advanced technologies, such as IoT sensors, block chain, and data analytics, enable real-time monitoring, traceability, and transparency throughout the supply chain. These innovations help identify potential risks, such as product spoilage, contamination, or transportation delays, allowing producers to implement proactive measures to mitigate or prevent disruptions. For example, IoT sensors can monitor temperature and humidity levels in storage facilities, while block chain technology can track the movement of products from farm to fork, providing assurance of product quality and safety. Furthermore, collaborative partnerships along the supply chain enhance risk management capabilities. By engaging with suppliers, distributors, logistics providers, and other stakeholders, producers can share knowledge, resources, and best practices to collectively address supply chain risks. Collaborative initiatives, such as supply chain mapping, contingency planning, and joint risk assessments, enable stakeholders to identify vulnerabilities and develop strategies to minimize the impact of potential disruptions on market access and product delivery.

Therefore, the effective risk management is crucial for ensuring market access and driving supply chain innovation for agricultural produce. By diversifying market channels, leveraging technology, and fostering collaborative partnerships, producers can mitigate risks, enhance resilience, and capitalize on opportunities in an increasingly dynamic and uncertain market environment. Prioritizing risk management strategies empowers agricultural producers to navigate challenges, optimize supply chain performance, and sustainably manage their operations to meet the demands of consumers and markets.

Collaborative Partnerships

Collaborative partnerships play a vital role in enhancing market access and driving supply chain innovation for agricultural produce. By fostering cooperation and coordination among stakeholders across the agricultural value chain, collaborative partnerships enable producers to overcome barriers, leverage resources, and unlock new opportunities in domestic and international markets. One key benefit of collaborative partnerships is the pooling of resources and expertise to address common challenges. Agricultural producers, processors, distributors, retailers, government agencies, research institutions, and non-governmental organizations (NGOs) can come together to share knowledge, technology, and best practices, enabling collective problem-solving and innovation. For example, collaborative partnerships can facilitate the adoption of sustainable farming practices, improve post-harvest handling techniques, or develop market intelligence systems, thereby enhancing the competitiveness and sustainability of agricultural supply chains.

Moreover, collaborative partnerships enable stakeholders to leverage each other's networks and market access to reach new customers and expand market reach. By forming strategic alliances and joint ventures, producers can access distribution channels, retail outlets, and export markets that may have been previously inaccessible. For instance, partnerships between agricultural cooperatives and export firms can facilitate market entry into overseas markets, while collaborations with e-commerce platforms can enable direct-to-consumer sales and market penetration in urban areas. Furthermore, collaborative partnerships foster innovation and technology adoption throughout the supply chain. By collaborating with technology providers, startups, and research institutions, agricultural producers can access cutting-edge technologies, such as IoT sensors, block chain, and precision agriculture tools, to improve productivity, quality, and efficiency. Collaborative innovation ecosystems enable rapid prototyping, testing, and scaling of new solutions, driving continuous improvement and competitiveness in agricultural supply chains. In conclusion, collaborative partnerships are essential for enhancing market access and driving supply chain innovation in the agricultural sector. By fostering cooperation, sharing resources, and leveraging collective expertise, stakeholders can address common challenges, access new markets, and adopt innovative solutions to improve productivity, sustainability, and competitiveness. Prioritizing collaborative partnerships enables agricultural producers to navigate complex market environments, seize opportunities, and create value for all stakeholders along the agricultural value chain.

Conclusion

The market access and supply chain innovation are paramount for the agricultural sector's sustainable growth and resilience in the face of evolving challenges and opportunities. Market access ensures that agricultural producers can effectively reach consumers and capitalize on market demand, while supply chain innovation optimizes the processes involved in getting products from farm to market. Through diversification of market channels, adoption of digital technologies, and fostering collaborative partnerships, agricultural producers can overcome barriers to entry, reduce risks, and unlock new opportunities in both domestic and international markets. Diversification allows producers to spread their risks across different markets, mitigating the impact of market fluctuations or disruptions in any single market. Meanwhile, digital technologies such as e-commerce platforms, block chain, and IoT offer unprecedented opportunities for enhancing transparency, traceability, and efficiency in agricultural supply chains. Collaborative partnerships bring together stakeholders along the agricultural value chain, enabling knowledge-sharing, resource pooling, and collective problem-solving. By collaborating with research institutions, technology providers, government agencies, and NGOs, agricultural producers can access expertise, funding, and innovative solutions to address common challenges such as food safety, sustainability, and market access. Furthermore, supply chain innovation, including infrastructure development, adoption of sustainable practices, and implementation of advanced technologies, improves the efficiency, resilience, and sustainability of agricultural supply chains. Investments in transportation networks, storage facilities, and processing plants facilitate the timely delivery of agricultural products to markets, reducing post-harvest losses and improving market access. Sustainable practices, such as organic farming, regenerative agriculture, and fair trade initiatives, not only meet consumer demand for ethical and environmentally friendly products but also enhance the long-term viability of agricultural production. In summary, market access and supply chain innovation are critical drivers of agricultural development, enabling producers to navigate complex market environments, optimize resource utilization, and meet the evolving needs of consumers and markets. By embracing market access strategies and supply chain innovations, the agricultural sector can contribute to economic growth, food security, and environmental sustainability, ensuring a prosperous future for both producers and consumers worldwide.

References

Ali, J. 2016. Adoption of innovative agricultural practices across the vegetable supply chain. International Journal of Vegetable Science, 22(1): 14-23.

Anderson, J.C. and Narus, J.A. 1991. Partnering as a focused market strategy. California Management Review, 33(3): 95-113.

Bardi, E.J., Raghunathan, T.S. and Bagchi, P.K. 1994. Logistics information systems: The strategic role of top management. Journal of Business Logistics, 15(1): 71.

Das Nair, R. and Landani, N. 2020. Making agricultural value chains more inclusive through technology and innovation (No. 2020/38). WIDER working paper.

Devaux, A., Torero, M., Donovan, J. and Horton, D. 2018. Agricultural innovation and inclusive value-chain development: a review. Journal of Agribusiness in Developing and Emerging Economies, 8(1): 99-123.

Ethier, W.J. 2007. The theory of trade policy and trade agreements: A critique. European Journal of Political Economy, 23(3): 605-623.

Kadaba, D.M.K., Aithal, P.S. and KRS, S. 2023. Impact of Aatmanirbharta (Self-reliance) Agriculture and Sustainable Farming for the 21st Century to Achieve Sustainable Growth. Mahesh, KM, Aithal, PS & Sharma, KRS (2023). Impact of Aatmanirbharta (Self-reliance) Agriculture and Sustainable Farming for the 21st Century to Achieve Sustainable Growth. International Journal of Applied Engineering and Management Letters (IJAEML), 7(2): 175-190.

Kalamkar, S.S., Ahir, K. and Bhaiya, S.R. 2019. Electronic National Agricultural Market (eNAM) in Gujarat: Review of Performance and Prospects. AERC Report, 177.

Ketikidis, P.H., Koh, S.C.L., Dimitriadis, N., Gunasekaran, A. and Kehajova, M. 2008. The use of information systems for logistics and supply chain management in South East Europe: Current status and future direction. Omega, 36(4): 592-599.

Kumar, R., Singh, R.K. and Shankar, R. 2016. Study on collaboration and information sharing practices for SCM in Indian SMEs. International Journal of Business Information Systems, 22(4): 455-475.

Rodrik, D. 2018. What do trade agreements really do?. Journal of economic perspectives, 32(2): 73-90.

Roy, D. 2022. Agriculture Marketing in India: Perspectives on Reforms and Doubling Farmers' Income. Journal of Marketing Development & Competitiveness, 16(3).

Seth, A.N.K.U.R. and Ganguly, K.A.V.E.R.Y. 2017. Digital technologies transforming Indian agriculture. The Global Innovation Index, 105-111.

Somashekhar, I.C., Raju, J.K. and Patil, H. 2014. Agriculture supply chain management: a scenario in India. Research Journal of social science and management, 4(07): 89-99.

Zilberman, D., Lu, L. and Reardon, T. 2019. Innovation-induced food supply chain design. Food Policy, 83: 289-297.

7

Adoption of Smart Irrigation Technologies

Hritul Kumar Gautam[1], Shashikant Gupta[2], Dhananjay Kumar[1], Manoj Gaund[1]

[1]Department of Fruit Sciences, College of Horticulture, Banda University of Agriculture and Technology, Banda, Uttar Pradesh
[2]Department of Agriculture, Horticulture, Deen Dayal Upadhyaya Gorakhpur University, Gorakhpur, Uttar Pradesh

Abstract

Countries are working together to use modern technology to enhance agricultural practices and increase efficiency. Therefore, increasing efficacy of irrigation in agriculture is essential to maintaining sustainable agricultural output. With the advent of wireless communication technologies, monitoring tools, and improved control strategies for effective irrigation scheduling, smart irrigation approaches can increase irrigation efficiency. The study evaluated many research topics in order to look into scientific methods for intelligent irrigation. This led to a broad range of themes being covered in this project that were associated with technology, decision-making, and irrigation techniques. Data was collected from several scientific publications. Thus, our study drew upon a number of published materials, the vast majority of which came out in the previous four years, and writers from around the globe. Special attention was paid to various irrigation activities in the meantime. After that, the assessment concentrates on the essential elements of smart irrigation, including energy harvesting, real-time irrigated scheduling, IoT, the significance of a web connection, and smart sensing. Nations are striving to improve agriculture's efficiency by using various technologies to make it more sustainable. Improving irrigation systems is essential for maximizing water usage efficiency and especially contributes to the United Nations' Sustainable Development Goals (SDGs). The purpose of this study is to demonstrate how smart irrigation, which makes use of sensor systems and the Internet of Things (IoT), contributes to the SDGs. The study focuses on secondary data gathering methods and has a qualitative methodology. Water conservation requires automated irrigation systems;

therefore, this advancement might be crucial in reducing water consumption.

Keywords: *Smart irrigation, Smart agriculture, IoT (Internet of Things), Sensors, Agricultural Machines*

1. **Introduction**

Agriculture is the backbone of the economy and a significant sector. For all nations, agriculture digitization is a growing area of concern. The need for food rises in tandem with the world's population growth, which is happening quite quickly. The agriculture sector is finding it very challenging to develop methods and strategies that will enable them to completely meet the growing wants and specifications due to the expanding need for food and shifting consumer preferences. Misuse of agricultural components can have detrimental effects on the environment and cause pollution. Farmers mostly rely on irrigation canal networks, groundwater, and rainfall in the field of irrigation. About 100 million of India's 500 million arable acres are irrigated with ground water; yet, 80% of this water is squandered owing to ineffective techniques. Reduced yield can be attributed in large part to over- and underwatering. Our goal is to create Internet of Things (IoT)-enabled irrigation systems that have the capacity to completely transform crop watering practices and significantly increase crop productivity, quality, and sustainability. Indian farmers may automate their irrigation procedures and make well-informed decisions to maximize water usage, conserve electricity, and extend the life of their machinery by incorporating actuators and sensor devices into an IoT network.

This paper provides an overview on Internet of Things smart irrigated systems in the country of India, including the most recent technology developments and key components. We also discuss the benefits of smart irrigation systems, including improved techniques for managing water, increased agricultural yields, and less water waste. Given the increasing worldwide need for food and freshwater resources, as well as India's considerable agricultural potential, the creation and deployment of IoT-driven automated irrigation systems becomes essential to attaining sustainable agriculture and guaranteeing food security in the nation [1]. Lack of field experience and the shortage of land reservoirs could be two of the main causes. Unirrigated land zones have developed as a result of the earth's water levels being lowered due to ongoing water extraction. As a result, developing agricultural systems has become essential, and nations are now trying to put in place efficient frameworks that will allow systems to function well [2].

An innovative method for automating irrigation systems and reducing water consumption is the smart irrigation system, which improves performance. This method enables farmers to fulfil their requirements with a newly accepted strategy that conserves water for the irrigation process by adjusting irrigation depending on actual land and weather conditions [3]. The demonstrates how wireless connection, data processing, fault detection, irrigation control, and data collecting (sensor) are all included in smart irrigation systems. These parts are all compatible with Internet of Things devices. Farmers can now have even more access to information on the precise state of their land, including soil temperature, water requirements, weather, and much more, thanks to technology like sensors, smartphone apps, and the Internet of Things (IoT) [4]. In light of the SDGs, this article seeks to demonstrate the value of smart irrigation employing IoT and sensing systems. This review will help farmers and researchers gain a better understanding of irrigation techniques and will offer a sufficient methodology for carrying out irrigation-related tasks [5].

2. Present difficulties and potential futures

This section addresses the opportunities and difficulties associated with deploying machine learning. The development of digital software and machine learning for innovative irrigation systems that manage different crops, particularly to support sustainable agriculture, has a number of challenges. To overcome the fold shortages, agricultural output must be raised overall. To meet demands from industry, more income crops like rubber and cotton must be planted, especially if they are combined with a sustainable crop to avoid damaging the soil. Additionally, numerous challenges are presented by these issues, such as the decrease in agricultural labor, the reduction in the amount of land that is suitable for cultivation, the lack of adequate drinking water supplies, the consequences of global warming, etc. The world's population is rapidly becoming older and moving away from rural areas in Favor of urban locations. There are several potential applications for IoT methods in irrigation systems in agriculture and food production. More focus is needed on a number of IoT-related aspects of smart irrigation, such as cost, dependability, robust design, portability, autonomous operation, and low maintenance. Agriculture is expected to transform into a dynamic business as integrated systems realize the potential of big data and artificial intelligence. These systems of integration will incorporate a range of agricultural implements, machinery, and management strategies that can be utilized for everything from plantings to yield prediction.

Advanced machinery including agricultural machines, the use of cloud computing, machine learning, and big data may usher in an entirely novel phase of IoT in the agricultural sector. In order to maintain sustainable agriculture, these instruments are seen to be very important. When farmers and other stakeholders combine machine learning forecasts with portable software solutions, there are numerous opportunities. Improving irrigation demand forecasts, coordinating volume and timing with plant requirements, and adaptively correcting for water loss are ways to increase the efficiency of water use. Improving irrigation demand forecasts, coordinating volume and timing with plant requirements, and adaptively correcting for water loss are ways to increase the efficiency of water use. This will use less irrigation water and provide a higher yield. A more sophisticated and intelligent model will be used as the system develops to make better irrigation decisions. As a result, farmers and users might experience a significant decrease in the stress and workload related to irrigation [6].

3. The Advancement of Irrigation

This section examines the evolution of irrigation in four distinct time periods, spanning from 1970 to 2022. Between 1970 and 1985, the onset of intelligence monitoring systems and irrigation water constraints piqued the interest of researchers in irrigation optimization. When water demand started to increase due to population development and the depletion of natural resources in the late 1970s, information and efficiency regarding water usage were introduced. The situation demanded that the irrigation method be improved. In order to achieve irrigation optimization, it was determined that the stress per day index (SDI), components of normalized crop susceptible (NCS), the transpiration of water, crop canopy growth and climate-related variables were all significant [7].

Following the public introduction of the Internet in 1989, there was a surge in the creation of web-based data storage and internet-based control systems

[8]. WSNs have started to gain popularity in 2000 as an easy-to-use and reliable environmental monitoring technology. Numerous WSN applications, particularly agricultural ones, have led to the development of actuators and sensors. By providing the grower with constant input regarding the crop's water requirements, WSNs add value to current irrigation systems. Additionally, build WSNs with effective routing protocols and the ability to monitor and control irrigation applications with water using a variety of techniques [9]. Using state-of-the-art advanced technologies such as the use of machine learning (ML), machine learning, artificial intelligence (AI), unmanned aerial vehicles (UAV), and the Internet of Things (IoT), researchers in precision agriculture have been closely examining innovative uses and approaches for irrigation, soil fertilization, insect management, and disease forecasting [10].

4. Importance of Water Management in Smart Irrigation

When it comes to irrigation, controlling water could be regarded as a crucial idea. Global concerns over clean water scarcity have led to a need for the agricultural industry and other industries to pay close attention to this problem. One way to think of water management is as controlling soil moisture to ensure that the proper amount and level of water is utilized when it's needed [11]. For the agriculture industry, efficient water management is vital since it can save costs and increase crop yield. The administration of water is also essential since it enables agriculture sector organizations to properly manage resources and complete necessary tasks. Given that diverse projects are being undertaken at varying scales, it is crucial to ascertain whether or not these initiatives will be completed successfully [12]. Ensuring the highest possible degree of job efficiency is only one of the many reasons why water management is so important. It might be able to ensure that agricultural crops receive the right amount of water in arid regions and during dry spells by using water management [13].

Since many projects are being implemented in places with limited water resources, attention must be paid to water management in order to distribute and apply water in a timely manner. Furthermore, rainfall is often quite insufficient in many regions of the world, thus a lot of water needs to be stored to ensure that the lack of rainfall can be compensated for. A few of the following could be named as follows:

- Quantify/measure/oversee
- Making use of water-wise irrigation and landscaping
- Manage reverse osmosis;

- Gather rainfall;
- Construct reservoirs

These could be considered some of the agricultural industries' most successful approaches to water management. Although these techniques could lead to efficient water management, a lot depends on how effectively they are implemented and how well the results can be inferred.

5. Why Is Need for Innovation Smart Irrigation Technology?

Many urgent issues with conventional irrigation techniques and the larger picture of contemporary agriculture necessitate development in smart irrigation technology. The following are some main justifications for why smart irrigation technology innovation is crucial:

1. Water Scarcity and Resource Management

a. Due to population expansion, climate change, and altered weather patterns, many parts of the world are experiencing water scarcity.

b. Water shortage challenges may be addressed by using smart irrigation technology to optimize water consumption and provide effective and targeted watering.

1. Growing Demand for Food Production

a. The demand for food production is rising in tandem with the world population growth.

b. By enhancing crop yields and quality, smart irrigation technology can assist farmers in satisfying the increasing demand for agricultural products without having to drastically increase the area under cultivation.

2. Environmental Sustainability

a. Excessive use of water sources and environmental deterioration are common outcomes of conventional irrigation techniques.

b. Intelligent irrigation systems minimize the negative effects of agriculture on the environment by promoting sustainable water management and cutting down on water waste.

3. Energy Efficiency

a. When it comes to pumping water for vast agricultural areas, traditional irrigation methods can be energy-intensive.

b. Smart irrigation solutions can help save energy and lessen carbon emission by adjusting watering schedules according to real-time data.

4. Precision Agriculture

a. A key element of precise agriculture, which uses technology to maximize numerous elements of farming, is smart irrigation.

b. Farmers can use resources like water, fertilizer, and pesticides more effectively thanks to precision agriculture, which boosts yields and has a smaller negative impact on the environment.

5. Economic Sustainability for Farmers:

a. By optimizing resource usage, lowering water and energy costs, and enhancing overall farm management, smart irrigation systems may help farmers save money.

b. Higher productivity and efficiency can improve agriculture's economic sustainability for the individual farmer and the sector as a whole.

6. Technology Integration and networking:

a. New avenues for incorporating automated irrigation systems into more comprehensive agricultural management systems have been made possible by developments in sensor technology, data analytics, and networking.

b. The use of technology enables automated control, real-time monitoring, and more accurate decision-making, resulting in more efficient and adaptable agricultural methods.

8. Adaptation to Climate Change:

a. Farmers find it difficult to anticipate the best times to irrigate their fields due to the unpredictability and fluctuation of weather patterns brought about by climate change.

b. Farmers can better adjust to changing climatic conditions by improving irrigated in response to both current and predicted weather changes thanks to smart irrigation technology, which include technologies like climate prediction and sensor-driven management.

6. Component of Novel Irrigation Technology:

6.1. Use of Sensors

It is frequently not feasible to upgrade to a smart irrigation controller when a property already has a scheduling controller in place. An improvement to the current system may involve adding a soil moisture, rain, wind, or freeze sensor to boost the effectiveness of automatic irrigation systems. A few manufacturers make instruments that can measure more than one environmental factor using a single apparatus. Numerous sensors are straightforward to install, work with current systems, and yield outcomes that are comparable to those of smart irrigation controllers. If an appropriate irrigation timer is already placed on the property, add-on sensors are often less expensive than smart irrigation controllers [14].

6.2. Sensors for Soil Moisture

It is possible to link soil moisture sensors to an existing irrigating controller. Before a planned watering event, the sensor determines the amount of moisture in the growing area of the soil. If the soil saturation is higher than a certain threshold, the cycle is skipped. There are several kinds of soil moisture detectors available, and before making a purchase, the customer should confirm that the sensor is compatible with their system. A soil freeze sensor, which stops a watering cycle if the temperature drops below 32 degrees Fahrenheit, is a feature of certain soil moisture sensors. Wireless or cable solutions are offered for soil moisture sensors. [15].

6.3. Sensors for Rain and Freeze

Rain and freeze detectors stop the irrigation cycle when it's not essential to water during a precipitation or freeze event, even if they're not classified as smart technology. Water, money, and needless runoff are all wasted when irrigation occurs during rainy seasons. There are three varieties of rain sensors accessible, and each one has a particular purpose based on a different idea [16].

6.4. Sensors of Wind

Oklahoma has gusts of 20 to 30 miles per hour (mph), with an average sustained wind speed of 16 mph. Watering in windy weather affects the quantity of water that seeps into the soil profile and the uniformity of irrigation distribution over the landscape. If wind speed rises over a predetermined level, wind sensors cut off the watering cycle. In addition to creating a beautiful, healthy environment, intelligent irrigation technology may assist minimize water waste? Owners of irrigation systems should make sure the system is regularly maintained and that the landscape is only watered when necessary [17].

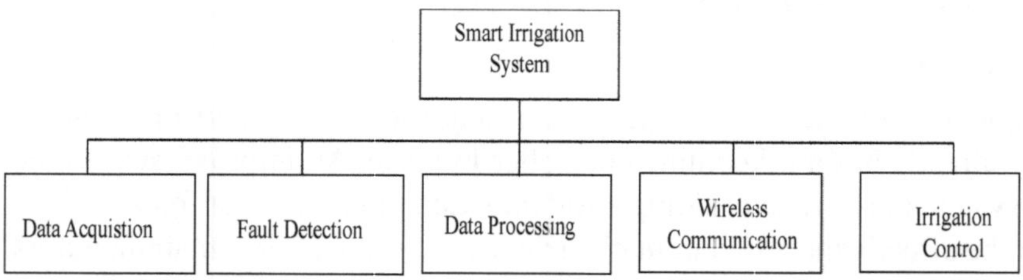

7. Soil and meteorological surveillance

Designing an irrigation management system that optimizes food production while minimizing water waste requires the use of effective and reliable monitoring systems, which have an impact on plant development and growth [18]. Using Internet of Things (also known as IoT) and wireless sensors networks (WSN), monitoring in the particular environment of precision irrigation involves gathering data that appropriately leads to represent the real-time state of the vegetation, soil, and climate of irrigation regions. The Internet of Things (IoT) has made it possible to create a low-cost technological solution that improves the management and oversight system for the process of irrigation while also establishing a real-time monitoring system. This method involves creating a network of nodes of wireless sensors that can perceive, process, and send data on various criteria.

Water-on-demand irrigation and suspended cycle irrigation are the two main soil moisture sensor-based irrigation systems. A suspension cycle is more like to a conventional timer controller, complete with irrigation, length, start, and finish time schedules. This is different in that when the soil reaches the proper moisture level, the device will automatically cease the next scheduled watering. However, there is no need for programming or a set amount of time for water-on-demand irrigation. With this method, the user sets the threshold, which triggers irrigation when the soil's moisture content falls below predetermined values. [19]

8. Internet of Things (IoT) and Smart Systems Used in Irrigation

8.1. Technologies of communication

The communication technologies that are being utilized in conjunction with the deployment of IoT devices may be seen as essential to achieving effective operations. It is also possible to see the use of communication technologies as being in line with the context in which they will be employed [20]. Two categories might be used to group the primary IoT technologies for irrigation. One might be thought of as the node-functioning devices that have minimal energy consumption, forward or transmit modest amounts of data over short distances. As a result, the other gadgets are the ones with high energy consumption that can send massive volumes of data across large distances. Wireless protocols are widely available for use in Internet of Things communication, and they may be used to categorize devices based on whether they communicate over long or short distances [21].

Wi-Fi has been found to be one of the most popular and efficient communication technologies because of its potential accessibility. It has also been found that the majority of the low-cost IoT devices now on the market support Wi-Fi, which is thought to be an effective general solution despite its limits in terms of area coverage and reach [22]. Furthermore, it has been determined that the Global System for Cell phone connection (GSM) is a widely used wireless technology that offers long-range connection; all that is needed is a mobile plan from a service provider that operates in that specific location. The more recent development of Long Ranging (LoRa) and messaging telemetry transport is another noteworthy technology. Because of its extremely long range, LoRa is a very viable technology that can be especially helpful in remote locations without service. Nonetheless, because of its minimal overhead and power consumption, MQTT has also become a commonly used protocol; nonetheless, irrigation systems are not yet using it extensively [23].

8.2. Cloud-based technologies

Cloud and conventional databases might be considered two of the more important and widely utilized storage systems. Because they enable the ability to store and retrieve important information when needed, these storage facilities are essential for a wide range of organizations operating in diverse sectors and industries. These kinds of storage solutions have facilitated the emergence of the big data idea, which describes vast quantities of information utilized by businesses for a variety of objectives [24].

Taking irrigation into account, cloud computing has also been utilized to provide warnings by using algorithms. By using these notifications, several dangers and hazards that would have otherwise occurred have been reduced. By using these notifications, it is easier to modify work schedules and take the required safety measures to lessen the likelihood of any difficulties [25]. Numerous cloud-related programmes have been developed to aid in job performance. Although each program has its own purpose and importance, its implementation is based on a number of criteria, including cost, services, adaptability, and other considerations. Given the hazards, losses, and complexities involved, maintaining the irrigation system might be viewed as a complicated task. Having stated that, those engaged in irrigation operations employ cloud technology to lower risks and enhance work output in order to meet predetermined goals [26].

8.3. IoT system advantages for irrigation

IoT systems for irrigation provide a number of advantages, some of which include reduced total water use, high cost and performance efficiency, reduced energy consumption, reduced crop waste, and more [27].

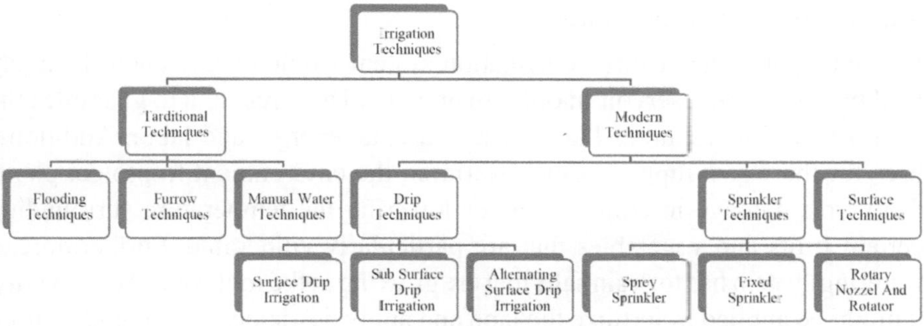

Smart irrigation uses water resources just to the level that they are needed, with little or no human intervention. Moreover, it has a high cost-efficiency since it uses less water and requires less accuracy during the process, which lowers total costs. Because the method requires equipment to run for shorter periods of time and incorporates planned breaks into the process, it also dramatically reduces total energy use. Furthermore, as firms must restrict their expenses to a certain degree due to resource limitations, cost containment and resource conservation are essential. When smart irrigation is used, the cost aspect is taken into account, making it possible to carry out associated tasks effectively for less money 30. Finally, an additional benefit is that more efficient irrigation and water management mean that plants and crops only receive the necessary quantity of water, which minimizes crop waste from either insufficient or excessive watering [28].

9. The Irrigation System's Method

Numerous sources of water may be gathered and utilized for a range of irrigation techniques. To guarantee that every plant gets enough water, the ultimate objective is to disperse water equally over the whole area. The purpose of contemporary irrigation systems is to directly feed water to the root zone of crops. Contemporary techniques effectively minimize water wastage, evenly disperse the supplied water and energy, and effectively oversee the irrigation stage.

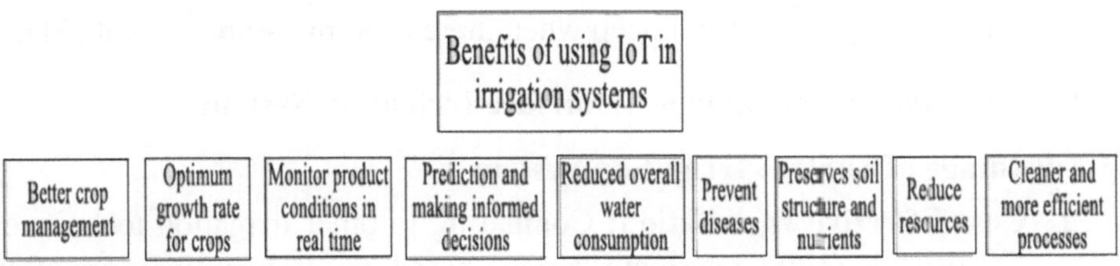

9.1. Surface Irrigation System

It is anticipated that a surface irrigation system would evenly and effectively feed the root zone reservoir in order to prevent plant stress and to guarantee the conservation of resources like water, nutrients, energy, and labor. Additional uses for the water supply include warming the environment to protect plants from frost damage in cold regions or lowering the temperature surrounding certain fruits and vegetables that are particularly vulnerable. Furthermore, a watering system has to drain salts that are growing in the root zone. Additionally, it might be applied to fertilize the land and apply pesticides, or it could be used to lighten the soil as preparation for improved farming. [29]. Surface irrigation is a widely utilized technique of irrigation that is quite prevalent due to its ease of use and low energy consumption. While inefficient irrigation application efficiency is typically linked to deep percolation and uneven irrigation water distribution, some research attempts to address these issues to improve the efficiency of surface irrigation [30].

The effectiveness of surface irrigation was enhanced to 86.6% by this study. A wirelessly connection that connects the sensors that monitor soil moisture and an auto checkpoint—which can be remotely operated via data from actual time soil moisture conditions—was used to construct and test an Internet of Things (IoT)-based system in the configuration of a level reservoir with a fixed point in sandy loam soil. An attempt was made to position the sensor in most advantageous location inside the basin plan in order to increase irrigation efficiency. In the water distribution centralization, an automated check gate made of aluminium with a steel structure was built to regulate the flow of water. Three capacitance-based soil moisture sensors were positioned at 25%, 50%, and 75% of the field's length, respectively, at depths of 37.5, 15, and 7.5 cm. Based on where the soil moisture detectors are located and the percentages of soil moisture deficit that were studied, there are three different operating schedules. According to the study, sensors in situations of soil shortage should be positioned at a depth of 37.5 cm and a distance of 25% from the injector when there is more moisture present. Sensors must be positioned 75% of the way from the entry and 7.5 cm deep when there is no moisture present [31].

Advantage and Dis-advantage of Surface Irrigation System

- **Advantage of Surface Irrigation System**

  1. **Cost-Effective Installation:** Comparing to other irrigation techniques like drip or sprinkler systems, surface-based irrigation systems are often less expensive to install. They are therefore a sensible option, particularly for farmers with tight budgets.

2. **Easy to Operate:** Managing and operating surface irrigation systems is not too difficult. Farmers don't need to be very skilled in technology to comprehend and use these systems.

3. **Suitability for a Range of Crops:** Row crops, cereals, and pasture crops are just a few of the many crops that surface irrigation is appropriate for. It is a recommended option in various agricultural applications because of its flexibility.

4. **Increased Soil Fertility:** Over time, increased soil fertility can be attributed to the water delivered by surface irrigation systems, which helps disperse nutrients in the soil.

5. **Lower Energy Consumption:** Because surface irrigation systems use gravity to spread water over the field, they frequently use less energy than pressured irrigation systems.

- **Disadvantages of Surface Irrigation System**

1. **Water Wastage:** The possibility of water waste is one of the main disadvantages of surface irrigation. Water resources may be used inefficiently as a result of evaporation, runoff, and infiltration.

2. **Uneven Water Distribution:** Using surface irrigation, it might be difficult to achieve consistent water distribution throughout the field. While some regions may receive little water, resulting in under-irrigation, other areas may receive excessive water, causing waterlogging.

3. **Land Requirement:** Because surface irrigation systems depend on the field's natural water flow, they frequently need a sizable quantity of land. For farms with a limited amount of space, this could be a constraint.

4. **Soil Erosion:** During surface irrigation, the flow of water over the field may exacerbate soil erosion. The quality and fertility of the soil may be impacted if this erosion results in the loss of priceless topsoil.

5. **Effort-intensive:** The surface irrigation networks may need physical effort for tasks like maintaining and preparing furrows. This may be detrimental, especially in areas where labor is costly or in short supply.

9.2. Drip Irrigation

One essential strategy for addressing the water shortage in the globe is drip irrigation. Drip irrigation is also known as trickle irrigation. Water is applied to plant roots using drip irrigation, which distributes water to the plant's roots drop by drop. This kind of irrigation may be among the most water-efficient as it minimizes evaporation and runoff. Drip irrigation is commonly employed in contemporary agriculture in combination with inorganic as well as organic (plastic) mulching, which offer further advantages like decreased evaporation, raised soil temperature, weed control, etc. On the various hand, the problem of drip pump emitter blockage significantly affects irrigation efficiency and uniformity; it can even result in the system being deactivated and lower crop output.

This study offers a drip irrigation system that is automated. For three months, the technique is tested in a paddy field. According to the testing setup, it saves around 41.5% and 13% of the water compared to traditional flood & drip irrigation systems, respectively. [32].

Advantage and Dis-advantage of Drip Irrigation System

- **Advantages of Drip Irrigation System**
  1. **Water Efficiency:** Because drip irrigation minimizes runoff and evaporation by delivering resource directly to the plant's root region of plants, it is incredibly water-efficient. With this focused approach, less water is wasted and the total quantity of water used can be precisely controlled.
  2. **Better Crop Productivity and Quality:** Better crop yields and better crop quality may be achieved with a continuous and regulated water supply and the capacity to manage the nutrients in the water. The proper amounts of water and nutrients are given to plants.
  3. **Decreased Weed Growth:** By directing water to the plants root zone directly, drip irrigation systems reduce the amount of moisture in the spaces between rows. Given that weed are more likely to flourish in drier environments, this may help inhibit the growth of weeds.
  4. **Energy Savings:** In general, drip irrigation systems use less energy than conventional surface watering techniques. This is due to the fact that they frequently run at a lower pressure and might not require pumping systems to distribute water.
  5. **Adaptability to Different Terrains:** Drip irrigation works well in a variety of environments, such as steep or uneven terrain. Because of its adaptability, it may be used in a variety of agricultural environments.

- **Disadvantages of Drip Irrigation System**

1. **Initial Cost:** Compared to conventional surface watering systems, installing an irrigation system with drips may be more expensive initially. Supplies, emitting devices, filtration systems, and control equipment costs are included in this.

2. **Clogging Problems:** Drip irrigation systems can get clogged, particularly in places where the water has a lot of silt or minerals in it. To keep the emitters from becoming clogged, regular maintenance is required, which includes the usage of filters.

3. **Skill Requirement:** Technical knowledge and skills are necessary for the proper design and implementation of a system for drip irrigation. For the system to be properly set up and managed, farmers could require training.

4. **Dependency on Stable Water Source:** The success of drip irrigation depends on a steady and stable water source. This can be problematic and call for more water storage options in places where water supply is erratic or irregular.

5. **Sensitive to External Factors:** Physical damage, differences in pressure, and temperature changes may all have an impact on drip systems. The effectiveness of the system may be impacted by severe weather or unintentional damage to its components.

9.3. Sprinkler Irrigation

Water is sprayed into the surrounding atmosphere and falls in a manner akin to rainfall in sprinkler irrigation. The water pressure regulates the spray of water, which travels via a system of pipelines and emerges through small nozzles. Carefully choosing nozzle diameters is important, taking into account the operating pressure and sprinkler formatting. The amount of water needed to irrigate crops and replenish the root zone can be utilized nearly evenly and at an acceptable pace, depending on how quickly soil leaks.

Numerous crops, including vegetables like onions, potatoes, carrots, cloves of garlic, lettuce and other varieties and others; spice like cardamom and pepper; flower like carnations and jasmine; oilseeds like sunflower, groundnut, and safflower; and fibres like cotton and sisal, may be cultivated using the spray watering system. Sprinkler irrigation works well on most soil types, with the exception of very clay soil. In addition, it saves water and gives the system mobility. Suitable for irrigating plants in areas with large plant populations per unit area; ideal for oil seeds and vegetables. Based on how portable they are, sprinkler irrigation systems come in a variety of forms, including fully lightweight, semi-portable, semipermanent, and totally permanent models.

Advantage and Dis-advantage of sprinkler Irrigation System

- **Advantages of Sprinkler Irrigation System**

 1. **Uniform Water Distribution:** Throughout the whole irrigated area, sprinkler systems offer a comparatively uniform water distribution. This promotes equitable growth and development of crops by guaranteeing that every plant receives the same quantity of water.

 2. **Variability in Crop Types:** Fruits, vegetables, and field crops are just a few of the many crops that sprinkler irrigation is appropriate for. It is a flexible option for a variety of agricultural applications due to its versatility in application.

 3. **Frost Protection:** Sprinkler systems have the ability to shield plants from extremely low temperatures by covering them in a sheet of ice. This is especially helpful for crops that are susceptible to frost.

 4. **Decreased Soil Erosion:** Because sprinkler systems apply water in a regulated manner, they can assist prevent soil erosion more than surface irrigation techniques. This is particularly helpful in areas with slopes.

 5. Labor Savings as Sprinkler systems may function automatically after installation, which lessens the requirement for labor-intensive physical work during irrigation. Increased operational efficiency and labor savings are possible outcomes of this automation.

- **Disadvantages of Sprinkler Irrigation System**

 1. **Water Loss from Evaporate and Wind Drift:** In windy weather, sprinkler systems may lose water from evaporation and wind drift. This may result in waste and a decrease in the overall effectiveness of water application.

 2. **Costs of Infrastructure and Energy:** Installing a sprinkler system necessitates a large investment in sprinkler heads, pumps, and pipelines. Energy expenditures for pumping water are one of the operating costs, which might potentially be high.

 3. **Sensitive to Clogged dishwashers:** Sprinkler head and nozzles can get clogged, especially in places where the water is much mineralized or contains silt. In order to guarantee appropriate water distribution and prevent obstructions, routine cleaning and repair are required.

4. **Not Appropriate for All Crops:** Overhead watering may have detrimental effects on certain crops, particularly those that are prone to foliar diseases. Furthermore, excessive power from water droplets striking a plant might harm some crops.

5. **Restricted Usage in Wind Conditions:** Strong gusts can distort sprinkler distribution patterns, resulting in an unequal misting of water. A sprinkler system's efficiency may be hampered by wind, necessitating modifications or the use of other watering techniques.

10. Barries of Smart Irrigation

1. **Costs and Affordability**: Despite possible long-term savings, many farmers, particularly people with small or medium in size operations, might discover the upfront costs prohibitive for purchasing and establishing smart irrigation systems, which include monitoring, control, and communication infrastructure.

2. **Lack of Education and Awareness**: Farmers might not be aware of the features and advantages of smart irrigation systems. A lack of training and instruction about the upkeep and operation of these systems may be a factor in people's resistance to embracing new technology.

3. **Technological Complexity:** Smart irrigation systems may be complicated, which might be a barrier for farmers who are not very tech-savvy. Farmers could be reluctant to implement systems if they believe they are too complicated or difficult for them to operate without the necessary training.

4. **Restricted Information and Support Access**: Adoption may be hampered by a lack of trustworthy information on available technology and support services.

   It might be difficult for farmers to get advice on choosing, setting up, and maintaining smart irrigation systems.

5. **Water Rights and laws:** The deployment of smart irrigation technology might be impacted by the considerable regional variations in water rights and laws.

   Why Farmers thinking about using these systems may get uncertain due to ambiguities in water-use legislation or limitations.

6. **Data Security and Privacy Concerns**: These worries might be a hindrance. They relate to the security and privacy of the data that smart

irrigation systems gather. Farmers can be hesitant to divulge private information for fear of data abuse or illegal access.

7. **Limited Infrastructure and access:** Certain rural regions do not have the infrastructure or dependable internet access required for smart irrigation systems.
8. In distant agricultural areas, limited access to electrical sources for operating sensors and controllers might provide difficulties.
9. **Fragmented Agricultural Sector:** Managing and executing large-scale smart irrigation projects may be difficult in such a dispersed environment, which is generally typified by the presence of several small-scale businesses and individual farmers.

Conclusion

In order to increase the effectiveness of irrigation in smart agriculture, this article reviews intelligent irrigation management and monitoring techniques. The scheduling and management of irrigation using monitoring techniques has been the foundation of this work. Additionally, a conversation on potential avenues for future research depending on research gaps has been planned. This relationship highlights the need for research in open fields utilizing a combination of soil-based cultivation, weather-based, and a plant-based monitoring tools along with a discrete forecast control system. Open area agricultural-irrigation systems, in contradiction to environmentally friendly controlled agriculture research, have unknowns that need to be explored. Therefore, the invention of process dynamics methods for irrigation systems and the effects of sophisticated management and monitoring strategies on watering yield in open field systems for farming will be the main areas of study in the future. In the current economic climate, technological advances have become indispensable for enterprises, and companies across all sectors are implementing upgrades in order to prosper and grow. This means that irrigation and the ways it is used may be enhanced to achieve the required performance outcomes with the highest possible operational efficiency. The idea of dealing with water has emerged, and businesses are drawn to finding ways to preserve the resource while simultaneously increasing productivity. In the modern business world, where firms are using technology to meet their performance targets, Smart irrigation systems have become essential. The implications of sensor systems and the Internet of Things have been critical. IoT lowers the overall cost of technology, making it possible to control irrigation process monitor systems. Another tool for real-time irrigation and precision farming monitoring is the wireless sensor network (WSN).

Reference

1.A. Clemmens, 1992. Feedback control of basin-irrigation system, Journal of irrigation and drainage engineering, vol. 118, no. 3, pp. 480–496.

2.A. Fahmi, 2020. Advanced internet of things irrigation mechanism, Int. J. Eng. Res. 9(1): 10-20.

3.A. Goap, D. Sharma, A. K. Shukla, and C. R. Krishna, 2018. An IoT based smart irrigation management system using machine learning and open-source technologies, Computers and electronics in agriculture, vol. 155, pp. 41-49,

4.A. Goap, D. Sharma, A.K. Shukla, and C.R. Krishna, 2018. An IoT based smart irrigation management system using Machine learning and open-source technologies, Computer. Electron. Agric. 155:41–49.

5.A. Humpherys and H. Fisher, 1995. Water sensor feedback control system for surface irrigation, Applied engineering in Agriculture, vol. 11(1): pp.61–65.

6.A.T. Abagissa, A. Behura, and S.K. Pani, 2018. IoT based smart agricultural device controlling system, in: 2018 Second International Conference on Inventive Communication and Computational Technologies (ICICCT), IEEE, , pp. 26–30.

7.C. Kamienski, et al., 2018. Swamp: an IoT based smart water management platform for precision irrigation in agriculture, in: Global Internet of Things Summit (GIoTS), IEEE, pp. 1–6.

8.C. Kamienski, et al., 2019. Smart water management platform: IoT-based precision irrigation for agriculture, Sensors 19 (2): 276.

9.Cardenas Lailhacar, B., M. D. Dukes, and G. L. Miller. 2010. Sensor-based automation of irrigation on bermudagrass, during dry weather conditions. Journal of Irrigation and Drainage Engineering. 136(3): 184-193.

10.Devitt, D. A., K. Carstensen, and R. L. Morris. 2008. Residential water savings associated with satellite-based ET irrigation controllers. Journal of Irrigation and Drainage Engineering. 134(1): 74-82.

11.E. Hussein Bani-Hani, M. El Haj Assad, M. Al Mallahi, Z. Almuqahwi, M. Meraj, and M. Azhar, 2022. Overview of the effect of aggregates from recycled materials on thermal and physical properties of concrete, Clean. Mater. 4.

12.F. S. Zazueta, A. G. Smajstrla, and G. A. Clark, 1994. Irrigation system controllers. University of Florida Cooperative Extension Service.

13.Food and A. O. (FAO). (2020) the practice of irrigation. [Online]. Available: http://wwwr.. fao.org/3/y3918e/y3918e10.htm

14.Gotcher, M., Taghvaeian, S. and Moss, J.Q., 2014. Smart irrigation technology: controllers and sensors. Oklahoma Cooperative Extension Service.

15.J. Knox, M. Kay, and E. Weatherhead, 2012. Water regulation, crop production, and agricultural water management Understanding farmer perspectives on irrigation efficiency, Agric. Water Manage. 108: 3–8.

16.K. Pernapati, IoT based low-cost smart irrigation system, in: 2018 Second Interna- tional Conference on Inventive Communication and Computational Technologies (ICICCT), IEEE, pp. 1312–1315.

17.L. Bortolini and M. Tolomio, 2019. Influence of irrigation frequency on radicchio (cichorium intybus l.) yield, Water, vol. 11: p. 2473.

18.L. García, L. Parra, J.M. Jimenez, J. Lloret, and P. Lorenz, 2020. IoT-based smart irrigation systems: an overview on the recent trends on sensors and IoT systems for irrigation in precision agriculture, 20 (4) 1042.

19.L.M. Fernández-Ahumada, J. Ramírez-Faz, M. Torres-Romero, and R. López-Luque, 2019. Proposal for the design of monitoring and operating irrigation networks based on IoT, cloud computing and free hardware technologies, Sensors 19 (10): 2318.
20.M. Ayaz, M. Ammad-Uddin, Z. Sharif, A. Mansour, and E.H.M. Aggoune, 2019. Internet-of-Things (IoT)-based smart agriculture: toward making the fields talk, IEEE
21.M. Monica, B. Yeshika, G. Abhishek, H. Sanjay, and S. Dasiga, 2017. IoT based control and automation of smart irrigation system: an automated irrigation system using sensors, GSM, Bluetooth and cloud technology, in: 2017 International Conference on Re- cent Innovations in Signal Processing and Embedded Systems (RISE), IEEE, pp. 601–607.
22.M. Soto-Garcia, P. Del-Amor-Saavedra, B. Martin-Gorriz, and V. Martínez-Alvarez, 2013. The role of information and communication technologies in the modernisation of water user associations' management, Comput. Electron. Agric. 98: 121–130.
23.M.S. Munir, I.S. Bajwa, and S.M. Cheema, 2019. An intelligent and secure smart watering system using fuzzy logic and blockchain, Comput. Electr. Eng. 77: 109-119.
24.Mayer, P.W. and Deoreo, W.B. 2010. Improving urban irrigation efficiency by using weather-based "smart" controllers. American Water Works Association. 102(2):86.
25.N. Sales, O. Remédios, A. Arsenio, Wireless sensor and actuator system for smart irrigation on the cloud, in: 2015 IEEE 2nd World Forum on Internet of Things (WF-IoT), IEEE, 2015, pp. 693–698.
26.R. Chen, H. Li, J. Wang, X. Guo, and Z. Song, 2021. Comparisons of spray characteristics between non-circular and circular nozzles with rotating sprinklers, Applied Engineering in Agriculture.
27.R.K. Kodali, M.S. Kuthada, and Y.K.Y. Borra, 2018. LoRa based smart irrigation system, in: 4th International Conference on Computing Communication and Automation (ICCCA), pp. 1–5,
28.S. Ghosh, S. Sayyed, K. Wani, M. Mhatre, and H. A. Hingoliwala, 2016. Smart irrigation: a smart drip irrigation system using cloud, android and data mining, in: 2016 IEEE International Conference on Advances in Electronics, Communication and Computer Technology (ICAECCT), IEEE, pp. 236–239.
29.S. Koduru, V.P.R. Padala, and P. Padala, 2019. Smart irrigation system using cloud and internet of things, in: Proceedings of 2nd International Conference on Communication, Computing and Networking, pp. 195–203.
30.T. R. Sinclair, C. Tanner, and J. Bennett, 1984. Water-use efficiency in crop production, Bioscience, vol. 34, no. 1, pp. 36–40,
31.V. Suma, 2021. Internet-of-Things (IoT) based smart agriculture in India - an overview, J. ISMAC 3: 1–15.
32.Z. Abedin et al., 2017. An interoperable IP based WSN for smart irrigation systems.

8

Skill Development and Agricultural Education

***Tripti Bhatia*[1], *Ankit Kumar Maurya*[2]**

[1]*Division of Animal Nutrition, College of Veterinary and Animal Science Bikaner, Rajasthan*

[2]*Agricultural Economics, Chandra Shekhar Azad University of Agriculture and Technology, Kanpur, U.P.*

Abstract

India's agricultural landscape faces significant challenges despite its abundant human resources and heavy reliance on agriculture for livelihoods. With a population of 1.39 billion, rural areas house over two-thirds of the total population, with approximately 58.3% of rural households depending on agriculture as their primary income source. However, low profitability and various obstacles such as low productivity, urbanization, fluctuating prices, and economic instability are driving farmers away from agriculture, contributing to a decline in the agricultural workforce. Skill development and agricultural education in agriculture emerges as a crucial solution to this crisis, aiming to enhance profitability, attractiveness, and entrepreneurship in farming. Empowering farmers with modern techniques and knowledge can replace traditional methods, making farming more efficient and economically viable. Furthermore, skill development initiatives can provide rural youth with self-employment opportunities, reducing rural-to-urban migration. This abstract highlights the urgent need to invest in skill development to ensure the sustainability and growth of India's agricultural sector amidst ongoing challenges.

Keywords: *Agricultural education, skill development, agriculture, self-employment, etc.*

Introduction

India possesses abundant human resources due to its demographic dividend. With a population of approximately 1.39 billion, rural areas are home to about

68.84% of the total population, constituting more than two-thirds (Census 2011, GoI). According to the National Sample Survey Office (NSSO) in 2019, about 58.3% of rural households rely on agricultural activities as their primary source of income. Despite the significant reliance on agriculture for livelihoods, the agricultural situation in India is challenging. Farmers are leaving agriculture, and young people are migrating to urban areas as farming becomes increasingly unprofitable. Issues such as low productivity, land and irrigation challenges, urbanization pressures, population growth, natural uncertainties, fluctuating prices, market risks, and economic instability exacerbate the situation. The rising number of farmer suicides underscores these challenges. Indian agriculture stands at a critical juncture, necessitating measures not only to make farming a lucrative pursuit to retain those engaged in it but also to boost agricultural productivity and income to address disguised unemployment in the agricultural workforce. Skill development in the agricultural sector emerges as a potential solution to this problem.

Need for Skill Development in Agriculture in India

The necessity for skill development in agriculture in India is becoming increasingly apparent as the agricultural workforce diminishes over time, largely due to farming's inability to provide profitable returns. According to the National Crime Records Bureau's (NCRB) annual report, more than 154 farmers and daily wage laborers commit suicide in India every day. Furthermore, the existing agricultural workforce earns only a fraction of the GDP, indicating that farmers earn less than a quarter of the average earnings of others.

Several indicators underscore the current crisis facing Indian agriculture

- The contribution of agriculture to India's GDP declined to 15% in 2022-2023 from 35% in 1990-91, largely due to rapid growth in the industrial and service sectors (GoI, 2023).
- Approximately 263 million people (54.6%) are employed in the agriculture sector, with over half of them now working as agricultural laborers (Census, 2011).
- The proportion of Indians employed in agriculture is dwindling. In 2021, 43.96% of the workforce in India were engaged in agriculture, down from 49.26% in 2011.

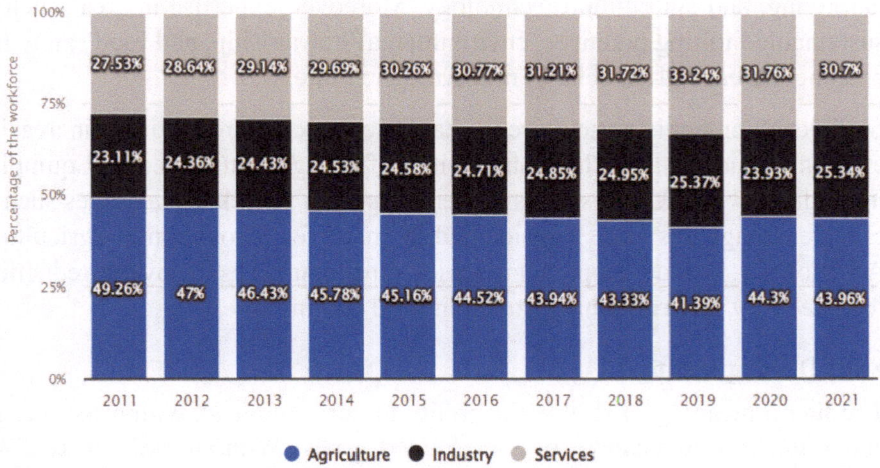

Therefore, it is imperative to enhance the profitability, attractiveness, and entrepreneurial aspects of agriculture to reduce rural-to-urban migration and instill pride in farming among farmers. To achieve this goal, it is essential to cultivate skills among farmers in various facets of farming, thereby replacing traditional, time-consuming, and costly methods with scientific, modern, economical, and efficient techniques. Empowering rural youth with farming and related enterprise skills enables them to pursue self-employment opportunities in their own villages with government support, instead of migrating to unfamiliar cities and leaving their families behind.

In India, the informal sector absorbs nearly 90 percent of the workforce, a substantial proportion of whom are either unskilled or possess inadequate skills, with scant provisions for formal skill development opportunities. Acquiring skill sets with tangible linkages to the labor market is indispensable for securing gainful employment, particularly within the informal sector. Skill, characterized as the capacity to proficiently perform tasks through training, experience, and practice, epitomizes skills development as a dual process involving the identification of skill gaps and the subsequent refinement and enhancement of these skills.

Importance of Skill Development in Agriculture

Agriculture's evolution has necessitated advancements in technology, practices, and techniques, subsequently altering the skill set required for success in the agricultural sector.

Farmers, agricultural workers, and professionals require a diverse set of skills, including crop production, livestock management, agribusiness management, marketing, and agricultural technology. Moreover, expertise in areas such as sustainable farming practices, environmental stewardship, and food safety has become increasingly crucial in modern agriculture.

Skill development in agriculture is vital for enhancing productivity, increasing efficiency, and ensuring the sustainability of the agricultural sector. Equipping individuals with the necessary skills enables them to tackle challenges such as climate change, resource scarcity, and the need for innovation in agriculture. Additionally, skill development in agriculture contributes to poverty reduction, food security, and economic growth in rural communities.

Skill Development

The literal meaning of skill is the ability to perform a task with effectiveness and efficiency to achieve pre-determined goals. Without skill even if we enough knowledge about a task, we may not be able to complete it with adequate dexterity. So developing skill in any job is the smart way of working. Skill or expertise can be acquired through training. Training is the process of organization of opportunities for participants to acquire necessary understanding and skill (Lynton and Pareek, 1990). Skill is the ability which is acquired through deliberate, systematic and sustained effort in order to smoothly and adaptively carryout complex activities involving idea s (cognitive skills) , things (t echnic a l skills), and/or people (interpersonal skills) (www.businessdictionary.com/definition/skill.html). In agriculture cognitive skills are required to make better decisions, technical skills required for handling various implements and interpersonal skills required for exchange of farm related information.

Need for Skill Development in Agriculture in India

The agricultural workforce over the years is gradually declining in India given the increasing failure of farming to ensure remunerative returns. It is assumed that India is losing 2000 farmers every day since 1999 (Sainath, 2013). Moreover whatever workforce is engaged in agriculture, survives on just a seventh of the GDP implying earning of farmer less than a fourth of what others do on an average (Gupta, 2015).

The following points just highlight the current crisis of Indian agriculture:

- 263 million people (54.6%) are engaged in the agriculture sector and over half of them are now agricultural labourers (Census, 2011) (Source: State of Indian Agriculture 2015-16, GoI).
- Total agricultural workers expected to decline to 190 million (2022) (expected decline of 33%) (IAMR, 2013).

In agriculture sector about 18.5% of the workers were skilled in 2009-10 (IAMR), of which less than 0.5 per cent have formal technical education Hence we can say that agriculture needs to be made more profitable, attractive and enterprising so that not only the rural to urban migration is reduced but also farmers start taking pride in their profession. For this we need to develop skills among our farmers in various aspects of farming so that the traditional, time and cost consuming methods are replaced by scientific, modern, economic and efficient methods. Rural youth once skilled in farming and related enterprises can choose self-employment in their own villages with Government help instead of migrating to unknown cities leaving their families behind.

What is Agricultural Education?

Agricultural education encompasses instruction on agriculture, food, and natural resources, imparting a diverse range of skills to students, including science, mathematics, communication, leadership, management, and technology. This education is disseminated through three interconnected components:

- Classroom or laboratory instruction.
- Experiential learning, which typically occurs outside the classroom under the supervision of agriculture instructors.
- Leadership education, facilitated through student organizations like the National FFA Organization, the National Young Farmer Education Association, and the National Postsecondary Agricultural Student Organization.

Many high school agriculture programs integrate FFA to bolster leadership and experiential learning. To delve deeper into FFA and its influence on agricultural education, visit www.ffa.org.

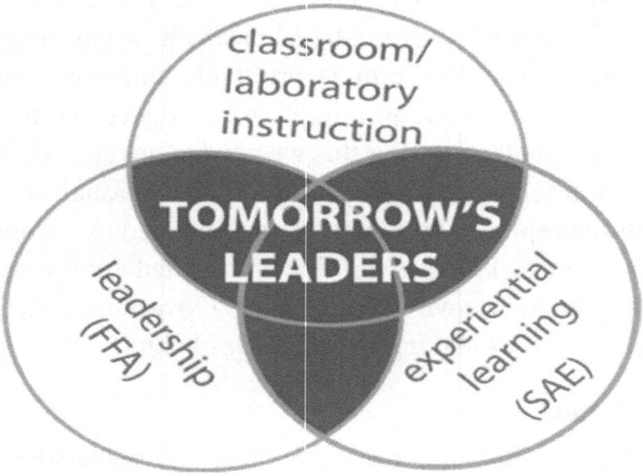

The Role of Agricultural Education

Agricultural education plays a pivotal role in skill development within the agricultural sector. It encompasses various formal and informal educational programs, including vocational training, technical schools, agricultural colleges, and university-level agricultural studies. Agricultural education not only provides individuals with technical knowledge and practical skills for agricultural production but also fosters critical thinking, problem-solving, and innovation within the industry.

Furthermore, agricultural education promotes entrepreneurship and prepares individuals for diverse career paths within agriculture, including farming, agribusiness, agricultural research, extension services, and agricultural policy. Through agricultural education, individuals gain an understanding of the complexities of modern agriculture and the inter connectedness of global food systems, which is crucial for addressing contemporary agricultural challenges.

Strategies for Effective Agricultural Skill Development

To ensure effective skill development in agriculture, it is essential to implement strategies that address the evolving needs of the agricultural sector. Key strategies include:

1. **Curriculum Development**: Educational institutions should continuously update their curricula to reflect the latest developments in agriculture, including technological advancements, sustainable practices, and market trends.
2. **Hands-On Training**: Practical, hands-on training is essential for developing technical skills in agriculture. This goal can be accomplished through internships, apprenticeships, and on-the-job training initiatives.

3. **Embracing Technology**: Integrating agricultural technology into educational programs can help students develop skills in precision agriculture, data analysis, and the use of agricultural machinery and equipment.
4. **Industry Partnerships**: Collaboration with agricultural industry stakeholders, such as farmers, agribusinesses, and research institutions, can provide students with real-world experiences and insights into the agricultural sector.
5. **Lifelong Learning**: Encouraging continuous learning and professional development among individuals already working in the agricultural sector is crucial for staying abreast of new developments and best practices.

National Policy on Skill Development

Mission

The National Skill Development Initiative aims to empower all individuals through improved skills, knowledge, and nationally and internationally recognized qualifications to gain access to decent employment and ensure India's competitiveness in the global market.

Aims

The aim of skill development in the country is to support achieving rapid and inclusive growth by enhancing individuals' employability, improving productivity and living standards, strengthening the country's competitiveness, and attracting investment in skill development.

Objectives

The objectives of the national policy on skill development include creating opportunities for all to acquire skills throughout life, promoting commitment by all stakeholders to own skill development initiatives, developing a high-quality skilled workforce/entrepreneur relevant to current and emerging employment market needs, enabling the establishment of flexible delivery mechanisms, and ensuring effective coordination between different ministries, the center, the states, and public & private providers.

Scope of National Policy on Skill Development

The National Policy on Skill Development encompasses a wide range of initiatives, including formal and informal apprenticeships, various types of training programs for self-employment and entrepreneurial development, adult education, retraining programs for retired or retiring workers, lifelong learning opportunities, non-formal training provided by civil society organizations, as well as e-learning, web-based learning, and distance education.

Skill Development Initiative Scheme

- National Skill Development Mission
- National Policy for Skill Development & Entrepreneurship
- Pradhan Mantri Kaushal Vikas Yojana (PMKVY)
- UDAAN
- STAR
- Vocationalization of Education

The importance of skill development initiatives in India's agriculture and food sector cannot be overstated, given the substantial proportion of the workforce employed within this domain. Acknowledging the pivotal role of skilled labour in enhancing productivity and income, policymakers have prioritized skill development through targeted policies and initiatives. These efforts aim to ensure the quality and relevance of training programs, emphasizing assessment and certification to meet the dynamic needs of the market.

Amidst these efforts, the partnership between government and private players emerges as a critical factor in implementing sustainable skill development programs. Collaboration and synergy between these stakeholders are essential to effectively address the multifaceted challenges facing skill development in agriculture.

Furthermore, the emergence of Agriculture 4.0 underscores the importance of digital education and ICT skills for farmers to adapt to technological advancements. As the industry embraces innovations such as robotics, drones, and IoT devices, farmers must acquire new skill sets to harness the full potential of these technologies and optimize their agricultural practices. However, despite India's rapid economic growth, persistent challenges remain in addressing skill gaps within the workforce, particularly in the context of the dominant informal sector. Bridging these gaps requires a concerted effort towards strengthening vocational education and training systems, ensuring alignment with market demands, and labour market linkages. In response to these challenges, the Skill India Mission has been launched with the ambitious goal of providing skill training to millions, with a particular focus on the agriculture sector. This initiative underscores the crucial role of skilled labour in driving sustainable development and economic growth within the agricultural domain.

In light of these complexities and opportunities, this chapter aims to explore the evolving landscape of agricultural education and skill development, with a focus on India's agriculture and food sector. By examining perceptions, policies, and initiatives, it aims to shed light on the pathways towards fostering a skilled workforce capable of meeting the evolving needs of the agricultural industry.

Given current population growth and demographic trends, with food demand expected to surge by 60%–70% by 2050, Asia requires another significant advancement in food production and transformative changes in agricultural and food systems. While the Green Revolution and other technological breakthroughs significantly enhanced agricultural productivity across the region, there is a pressing need to equip agricultural workforces with the skills required for the era of 'Industry 4.0'. However, at present, there is a notable gap in this regard. Thus, investing in education and skills development to create a new breed of 'agripreneurs' in Asian developing countries is crucial. Despite a significant share of the Indian population deriving their livelihood from agriculture, the Indian agricultural scenario is concerning. Farmers are leaving agriculture, while young people are migrating to urban areas as farming gradually loses its profitability. Indian agriculture grapples with challenges such as low productivity, land and irrigation problems, urbanization pressure, population growth, natural uncertainties, price fluctuations, market risks, and economic instability. The increasing number of farmer suicides is evidence of this fact. Indian agriculture is at a crossroads, and there is a pressing need to make farming a profitable venture to retain those who want to quit farming and enhance agricultural productivity and income to address disguised unemployment in the agricultural workforce of India. Skill development in the agricultural sector could be a possible solution to this problem. Upgrading skills development has been successful across all contexts and sectors, particularly in rural areas and the agriculture sector, as it yields positive impacts, especially on employment and earnings outcomes. However, this impact does not take effect immediately and is more pronounced in low- and middle-income countries than in high-income countries. Additionally, the effects vary greatly depending on the program type, design, and context, indicating the need for careful design of youth employment interventions. Evidence also suggests that targeting disadvantaged youth may be a key factor for success.

Effective skills development programs share several common features: They provide multiple pathways for learning and employment, focus on meeting employer and market demands for skills, utilize applied learning methods, and offer follow-up services and support to link farmers with tangible employment or self-employment opportunities. In the agricultural context, skills transfer effectively takes place in work-based learning settings such as farmer field schools, on-site employer-based training, internships, volunteer opportunities, and co-curricular youth organizations. Soft skills, such as social skills, positive self-concept, self-control, communication skills, and higher-order thinking, are as crucial to success in the workplace as technical or agricultural-specific skills. These soft skills are particularly important in the agriculture sector, where ever-changing global demands necessitate flexibility and adaptation. Evidence suggests that many of these skills, acquired in agricultural or workforce settings, are applied in other aspects of youth's lives, including conflict mitigation, health, and nutrition. These findings underscore the benefits of a cross-sectoral approach to youth development in agriculture.

Meanwhile, a lack of capital limits smallholder farmers' access to modern technologies, such as higher-yielding crop varieties and farm equipment, which could reduce the need for labour and enhance farm productivity. Throughout Asia, the migration of younger generations to urban areas and the shift to non-farm employment have left older generations as smallholder farmers. However, smallholder farmers are not yet fully prepared for the trend in agriculture towards the consolidation of production, processing, marketing, promotion, and distribution under integrated agro enterprises. There is a corresponding need to enhance and strengthen farmers' agricultural business knowledge, including business networks. These factors exacerbate the challenges smallholder farmers face in transitioning from low-productivity subsistence farming to higher-productivity enterprise farming.

Skill development initiatives in agricultural education in India

1. **Government Initiatives**: The Government of India has launched various skill development initiatives aimed at enhancing the skills of individuals engaged in agriculture and allied sectors. These initiatives may include skill development programs, vocational training, and capacity-building schemes targeted at farmers, agricultural labourers and rural youth.

2. **Agriculture Skill Council of India (ASCI):** Established in January 2013 under the Companies Act of the Ministry of Company Affairs, the Agriculture Skill Council of India (ASCI) aims to enhance capacity in agriculture and bridge the gap between laboratories and farms. ASCI seeks to upgrade the skills of cultivators, agricultural labourers, and workers engaged in organized and unorganized agriculture and allied industries. Its vision is to create a sustainable industry-aligned ecosystem for robust skill and entrepreneurship development in the Agriculture and Allied sector.

ASCI's objectives include

- Determining skills/competency standards and qualifications, and developing National Occupational Standards (NOS).
- Maintaining a skill inventory to facilitate individual choices.
- Developing sector-specific skill development plans and standardizing affiliation and accreditation processes.
- Affiliating, accrediting, assessing, and certifying Vocational Institutes/ Programs.
- Planning and executing Training of Trainers (ToT).
- Promoting academic excellence and establishing a well-structured, sector-specific Labor Market Information System (LMIS) to assist in planning and delivering training.
- Adopting global best practices.

3. Government Programs in Tandem: Pradhan Mantri Kaushal Vikas Yojana (PMKVY), launched on July 15, 2015, aims to skill India on a large scale with speed and high standards, driving towards the vision of a "Skilled India." Implemented by the National Skills Development Corporation (NSDC) under the Ministry of Skill Development and Entrepreneurship (MSDE), PMKVY is a flagship scheme facilitating skill development and aims to provide skill training to youth across various sectors, including agriculture. Under this scheme, training programs are conducted by affiliated training partners to impart industry-relevant skills and improve employability.

3.1 Attracting and Retaining Youth in Agriculture (ARYA), launched by ICAR, aims to impart entrepreneurial skills to rural youth through Krishi Vigyan Kendras (KVKs) to reduce youth migration by making farming a remunerative venture.

3.2 Krishi Vigyan Kendras (KVKs): Krishi Vigyan Kendras, established by the Indian Council of Agricultural Research (ICAR), play a significant role in agricultural extension and skill development at the grassroots level. These centers conduct training programs, demonstrations, and workshops to impart practical skills and knowledge to farmers and rural youth.

4. Higher Education Institutions: Agricultural universities and colleges in India offer degree programs and diploma courses in agriculture, horticulture, animal husbandry, and other related fields. These institutions often include practical training components to equip students with hands-on skills required for agricultural practices and entrepreneurship.

5. Private Sector Initiatives: Several private organizations and companies may also be involved in skill development initiatives in agriculture. These initiatives may include training programs, workshops, and capacity-building activities focused on specific agricultural technologies, practices, or value chains.

Knowledge-intensive agriculture

The emergence of knowledge-intensive agriculture in response to emerging challenges necessitates the utilization of technological innovations. This involves understanding and applying advanced information to improve agricultural production and marketing, efficiently manage risks, and sustainably increase productivity and profitability. Sophisticated application of information and communication technologies (ICT) in knowledge-intensive agriculture includes optimizing the development and adoption of modern varieties, farm production inputs and operations, and postharvest operations. High-level technologies have led to the development of genetically engineered crops with improved pest resistance, nanotechnology-formulated agrochemicals, and precision agriculture practices. Climate change has further spurred the development of "smart greenhouses" equipped with remote environmental sensors and other green technologies to automate microclimate control systems and manage weather, pests, and diseases.

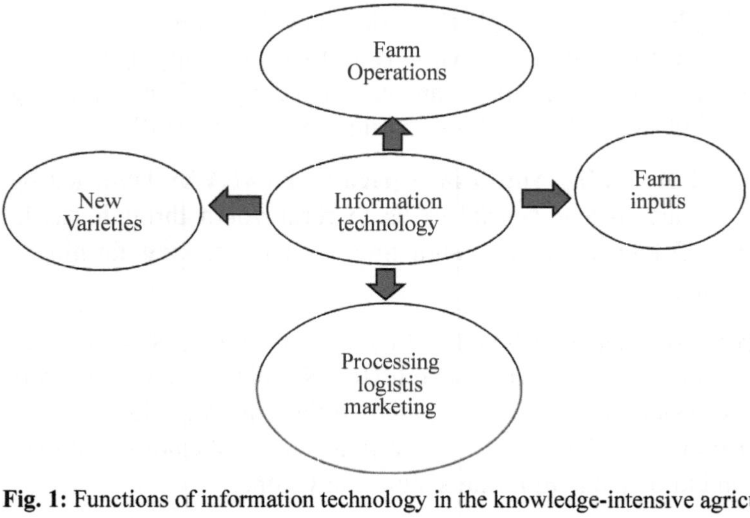

Fig. 1: Functions of information technology in the knowledge-intensive agriculture

In the People's Republic of China, farmers are employing drones for crop dusting and pesticide spraying over large areas within a shorter period and at lower costs for labour and equipment maintenance. Beyond the fields, high-level technologies encompass the entire food value chain, from new processing and packaging methods to smart logistics information systems and e-commerce or online marketing. "Internet-plus rural economy" or "digital agriculture" is also on the rise, leveraging internet technologies to boost efficiency in rural economies, diversify rural incomes, and enhance rural productivity and efficiency. Greater investment in agricultural innovation in the coming years, including agricultural ICT, biomaterials, and value chain and agribusiness initiatives, has the potential to further transform the sector.

Individuals involved in the agricultural sector, including farmers, farm managers, service providers, and processors, stand to benefit from acquiring new knowledge and skills through relevant education and training. These efforts are crucial for enhancing employability, boosting productivity, increasing earning capacity, and revitalizing the agriculture sector.

Education and skills development for agricultural transformation

In recent years, advancements in science and technology have brought about transformative changes across all sectors, including agriculture. However, while agricultural outputs have seen growth, the industrial and service sectors have experienced even faster expansion, resulting in a shift of employment away from agriculture to urban sectors offering better-paying jobs. Despite agriculture remaining the primary employer in many developing countries, the availability of skilled labour for the sector is limited. This shift away from agriculture poses challenges to intensifying labour required to increase agricultural outputs, particularly in densely populated Asian countries.

Education plays a pivotal role in addressing these challenges. Research indicates that farmers with higher levels of education or greater knowledge are better equipped to enhance farm operations, increase productivity, and improve profitability. However, it's noteworthy that some educated farmers may opt to leave agriculture for opportunities in other sectors offering better remuneration and working conditions, although their contributions through remittances to families in agriculture persist.

Various studies have explored the correlation between farmers' educational attainment and labour productivity, including their willingness to adopt new production technologies and improve farm management practices. For instance, a study conducted in the Philippines revealed that rice productivity could be boosted by 26% through education of the household head. In the context of knowledge-intensive technologies like drone technology and biotechnology, farmers need to absorb knowledge through modern communication technologies.

In India, younger farmers, particularly sensitive to income disparities between farm and nonfarm occupations, exhibit higher occupational mobility compared to their older counterparts. Higher levels of education and skill positively influence the decision to transition out of agriculture, contributing to the inadequacy of skilled labour in Indian agriculture due to the allure of higher-paying jobs in manufacturing and services sectors.

Revamping agricultural education

Revamping agricultural education is crucial to ensure that millions of smallholder farmers and agricultural labourers, each with unique skills and knowledge, are equipped to engage in knowledge-intensive agriculture. This necessitates updating curriculums and programs with the latest scientific knowledge and technological breakthroughs, enhancing institutional capacities, and fostering stronger links and partnerships between academia and industry across the agriculture value chain.

Knowledge-intensive agriculture demands farmers who comprehend sophisticated production technologies, modern farm management systems, and financial and logistical strategies. Establishing high-technology universities with adequate physical infrastructure and qualified human resources can facilitate the dissemination of these subjects, even to existing smallholder farmers, provided the programs are designed with a specific focus.

Strengthening the agricultural research and extension system

The agricultural research and extension system also requires strengthening to deliver the latest technologies and knowledge to farmers effectively. Agricultural extension services, traditionally focused on delivering agricultural inputs, need refurbishment to incorporate advanced technologies. For instance, e-extension holds promise in reaching farmers more efficiently through mobile communication and other innovative e-learning platforms.

Developing high - tech agripreneurs by investing in skills development and education

Investing in education and skills development is vital for nurturing a new generation of high-tech agripreneurs and supporting agriculture's transformation. Accessible and available advanced technologies, coupled with education, training, and skills development, can enhance employment opportunities, increase productivity, boost incomes, and facilitate agricultural transformation. Therefore, the role of agricultural education and skills development warrants thorough examination to foster the emergence of high-tech agripreneurs.

Over the last five decades, agricultural education and skills development programs have consistently contributed to agricultural development and economic growth in developing countries, although to a limited extent. However, shifts in employment demands driven by factors such as improved access to primary and secondary education, urbanization, and international markets have necessitated a reassessment of these programs. Vocational programs, whether secondary or postsecondary, can focus on providing direct training for farmers and farm labourers, while the role of higher education in agriculture is becoming increasingly significant. National governments, private sectors, and donor communities need to evaluate existing agricultural education and skills development systems and strengthen their relevance in workforce development and enhancing agricultural value chains based on lessons learned.

Reformed agricultural education and skills development systems should be functional, responsive to emerging challenges, gender-sensitive, and flexible. These systems must aim to enhance qualifications and competencies to improve employability in agricultural value chains, as well as to impart technical and business skills that meet evolving public and private sector needs. There should be a concerted effort to bolster the participation and contribution of female labour in high-technology agricultural activities through skills development, particularly as male agricultural workers seek better-paying nonfarm employment.

Moreover, these systems should offer short, flexible, convenient, and mobile skills enhancement programs to agricultural workers who may seek continuous education and training while working full time. Innovative, customized, and interactive training methods utilizing modern technology can help build skills, encourage participation, and enhance the productivity of elderly farmers and laborers in knowledge-intensive agriculture. Additionally, agricultural education and skills development systems can design or upgrade existing job-matching services to connect candidates, including elderly individuals, with suitable employment opportunities.

Increasing government investments

Despite the importance of investing in education, government expenditure on education as a percentage of gross domestic product (GDP) has remained stagnant in some Asian countries (ADB, 2018). While certain Pacific countries allocate a higher percentage of GDP to education, others like Armenia, Bangladesh, Sri Lanka, and Cambodia have invested less than 3% of GDP since 2000. In particular, government spending on secondary and postsecondary no tertiary vocational education as a percentage of total government spending is notably low across South Asia. Policymakers need to recognize the value of specific trade-related vocational training in enhancing employability, offering a more sustainable alternative to early school dropout, and supporting long-term workforce development.

Government investments are crucial for enhancing the quality of technical teachers, thereby improving the overall quality of education and skills development. Education and skills development systems should transition from a supply-oriented to a more market-demand-oriented training approach that aligns with evolving market demands. Innovations such as competency-based training and vocational qualification frameworks can establish appropriate quality standards in the workplace.

Harnessing public–private partnerships

Public-private partnerships (PPPs) are increasingly recognized as vital in knowledge-intensive agriculture. PPPs can address technical, financial, and other constraints in developing, promoting, and practicing knowledge-intensive agriculture. These partnerships range from simple procurement to full-blown collaboration, with private partners engaged in asset operation, service delivery, and technology uptake.

In the context of agricultural education and skills development, collaboration with the private sector can ensure that trainees are equipped with practical know-how and skills relevant to knowledge-intensive agriculture. PPPs can facilitate the integration of theory and practice and bridge the gap between rural communities and scientific knowledge, ultimately making agricultural education and skills development systems more responsive to market needs. In South Asia, governments and private institutions must collaborate to expand capacity for training, enhance the quality of workers, and facilitate knowledge transfer in the practice of knowledge-intensive agriculture (ADB, 2017).

Enhancing the policy environment

Governments also need to collaborate with other sector stakeholders to establish and uphold a conducive policy environment. Education and skills development systems should be integrated into countries' broader agricultural development strategies. The coordination among relevant institutions, including government bodies, training systems, local organizations, and donors, must be strengthened to support sector development priorities. These priorities include responding promptly to structural transformations, reducing rural unemployment or underemployment rates, and enhancing the productivity of rural labor (ADB, 2017).

Governments should pursue educational reform and advocate for lifelong learning (ADB, 2018). They ought to incentivize schools to strengthen foundational and other skills that enable individuals to engage in continuous learning. Universities and vocational training institutions will play crucial roles in imparting the specialized skills required to work with new technologies, thereby increasing the number of agripreneurs, including both new graduates and individuals seeking to update their skills.

Furthermore, governments can delegate the responsibility of skills training to the private sector, while ensuring that training remains accessible in areas where private options are either costly or unavailable, such as remote geographical regions (ADB, 2017). Governments should actively disseminate information to the public regarding current and future labour market demands, available training programs, and the performance of the public education system.

Scope for Skill Development in Agriculture for Farmers and Rural Youth

With increasing commercialization of farming, various new job opportunities are emerging in agriculture and allied sectors. Skill training can provide an advantage to those seeking job opportunities in these areas including:

- Agri Warehousing, Cold Chain, Logistics & Supply Chain,
- Digital Agriculture, weather forecasting
- Alternate energy (Solar, Biomass etc)
- Dairy, Poultry, Fisheries, Horticulture
- Farm Mechanization,
- Protected Cultivation (Green House, Hydroponics etc)
- Bamboo, Honey, Medicinal & Herbal

Implications of Skill Development

Skill development is crucial in agriculture to boost productivity and revitalize the workforce. It can address issues such as labour scarcity through farm mechanization, prevent rural migration by enhancing income and profitability in agriculture, strengthen the agriculture value chain, and equip the workforce for agricultural exports by ensuring product quality and meeting international standards.

Conclusions

Skill development and agricultural education are vital for ensuring the sustainability and prosperity of the agricultural sector. Asian developing countries must invest in education and skills development to foster entrepreneurship in the agriculture sector, cultivating what are often termed as high-tech agripreneurs. While productivity in the sector has improved in recent decades, another significant leap in food production is imperative, especially considering impending demographic changes. Demand for food in the region is projected to increase significantly by 2050, driven by population growth, urbanization, rising incomes, and shifting dietary preferences. It is crucial for Asia's farmers to transition from traditional input-driven agriculture to knowledge-intensive agriculture to meet this growing demand. Therefore, the time to revamp agricultural education and skills development is now. Advanced knowledge and technologies must become increasingly accessible to millions of smallholder farmers. This necessitates modernizing agricultural education and skills development systems. Universities can enhance curricula, institutional capacity, and industry partnerships to support the emergence of young, high-tech agripreneurs. Agricultural extension services can also transform existing agricultural workforces into agripreneurs by embracing e-learning to disseminate knowledge and technologies available under Industry 4.0. Agricultural education and skills development should be reformed into a functional, gender-sensitive, and flexible system that can effectively respond to emerging challenges. This transformation requires increased government investment, public-private partnerships, and an enabling policy environment. Government spending on education and skills development needs to be augmented in Asian countries to revamp agricultural education and skills development. The increased budget should be allocated for modernizing curricula, pedagogies, and laboratories, recruiting and developing quality teachers, and conducting industry-relevant research. Public-private partnerships in agricultural education and skills development, especially when aligned with a country's broader agricultural development strategy, can be instrumental in nurturing high-tech agripreneurs.

References

Amadi, N. S., & Nnodim, A. U. 2018. Role of agricultural education skills in entrepreneurship Development in Rivers State. International Journal of Innovative Social & Science Education Research, 6(1): 9-18.

Amadi, N.S. and Gibson, F.O. 2020. Role of agricultural education programme for entrepreneurship skill development among students in rivers state tertiary institutions. International Journal of Contemporary Academic Research, 1(1).

Bhattacharyya, S., & Mukherjee, A. 2019. Importance of skill development in Indian agriculture. ICT and social media for skill development in agriculture. Today & Tomorrow's Printers and Publishers, New Delhi, 47-62.

Brown, T. 2020. Pathways to agricultural skill development in the Indian Himalayas. Journal of South Asian Development, 15(2): 270-292.

Dailey, A.L., Conroy, C.A. and Shelley-Tolbert, C.A. 2001. Using agricultural education as the context to teach life skills. Journal of Agricultural Education, 42(1): 11-20.

Meghwal, P. K., Jadav, N. B., Reddy, S. Y., & Tripura, P. 2016. Skill development as a key for agriculture development.

Ra, S., Ahmed, M. and Teng, P.S. 2019. Creating high-tech 'agropreneurs' through education and skills development. International Journal of Training Research, 17: 41-53.

Ukonze, J. 2020. Innovative strategy for measuring skill performance of students of agricultural education for sustainable development in Nigeria. International Journal of Innovative Research and Advanced Studies, 6(7): 1-12.

9

Integration of Renewable Energy in Agriculture, Livestock and Poultry Farming

Tahir Manzoor[1*], Pankaj Kumar[2], Onkar Paradhe[3] and Kanta Godara[4]

[1]Department of Livestock Production Management, CVAS, Rajasthan University of Veterinary and Animal Sciences, Bikaner, Rajasthan*
[2]Department of Veterinary Physiology, CVAS, Rajasthan University of Veterinary and Animal Sciences, Bikaner, Rajasthan
[3]Department of Animal Nutrition, CVAS, Rajasthan University of Veterinary and Animal Sciences, Bikaner, Rajasthan
[4]Department of Livestock Production Management, CVAS, Rajasthan University of Veterinary and Animal Sciences, Bikaner, Rajasthan

Abstract

Renewable energy is derived from diverse self-renewing natural sources like sunlight, wind, water flow, geothermal heat, and biomass, reducing fossil fuel depletion, slowing climate change, and cutting carbon emissions. India ranks 3rd in global energy consumption and 4th in renewable energy installed capacity, with 179.3 GW installed by July 2023 a 396% increase in 8.5 years. Energy needs in livestock farming vary based on factors like type of farming and herd size, with electricity as a major and unavoidable expense. Renewable energy can power various farm operations including lighting, ventilation, heating, feed milling, milking, and product processing, etc. Solar, wind, biogas, ethanol, and biodiesel are key renewable energy types. Solar energy can be used to power water pumps, electric fences, milking machines, and electric tractors. Biogas, from anaerobic digestion of organic waste, and bio-slurry (potent fertilizer), offer sustainable options. India's 2018-2019 manure production (1700MMT) could supply 50% of LPG needs via biogas and meet 44% of NPK fertilizer requirements. Biogas also fuels generators for electricity. Biodiesel production reached 105.6 million Liters in 2019-20, while the Ethanol Blended Petrol (EBP) Programme, launched by the GOI in 2014, promotes ethanol. Wind turbines produces electricity from wind energy. The GOI initiated efforts to boost renewable energy production and usage through the establishment of the Ministry of New and Renewable Energy (MNRE) in 2006. Objectives include enhancing energy security, increasing renewable energy's share, and ensuring energy availability and affordability. Additionally, the GOI introduced schemes such as the New National Biogas & Manure Programme (NNBOMP) in 2018, Biogas Power Generation & Thermal Energy Application Programme (BPGTP) in 2019, and Waste to Energy Scheme (WTE) in 2021. Future prospects for renewable energy in livestock and poultry farming highlight the need for research, development, demonstration, deployment, and commercialization of new technologies, coupled with continuous surveys and feedback. Increased technological interventions in livestock farming will drive up energy demands, necessitating renewable energy-powered farms for fulfilment and increased profitability. Given growing climate change concerns, effective waste management is crucial, with biogas and bioenergy emerging as vital tools for mitigating climate impacts on the livestock industry.

Keywords: *Biodiesel, Bioenergy, and Renewable Energy, etc.*

Renewable Energy

Renewable energy, often referred to as clean energy, comes from natural sources or processes that are constantly replenished.

(National Research Development Corporation, India)

Renewable energy is energy derived from natural resources that replenish themselves over a period without depleting the Earth's resources.

(Centre for Resource Solutions, California)

The term "renewable energy" is energy derived from a broad spectrum of resources, all of which are based on self-renewing energy sources such as sunlight, wind, flowing water, the earth's internal heat, and biomass such as agricultural, industrial, and municipal waste.

(Bull, 2001)

Clean Energy

Clean energy refers to power generated by resources whose byproducts have minimal or no effect on the environment by reduction in emission of greenhouse gasses during production. Sun, wind, and hydropower are the main examples. This energy is considered clean because it doesn't produce the carbon dioxide and air pollution that come from fossil fuel consumption, which contribute to global warming.

Green Energy

Green energy refers to a subset of renewable energy sources that offer the greatest environmental benefit by having the smallest environmental footprint (no carbon emissions). These sources, including sunlight, wind energy, geothermal energy and hydro power.

To qualify for energy as green energy, a resource cannot produce pollution like fossil fuels. This clarifies that all energy sources used by the renewable energy industry may not be green. To give an example, power generation by burning organic material from sustainable forests may be renewable, but as it generates CO_2 it is not necessarily green energy.

Key Differences between Renewable, Clean and Green energy

1. Environmental Impact and Benefit

- **Renewable Energy:** The primary focus of renewable energy is on the sustainability and replenishing ability of energy sources over time. Environmental benefits are inherent in renewable energy, by reducing reliance on finite fossil fuels.
- **Clean Energy:** Focuses on energy sources that produce minimal or no harmful emissions during their production and use, encompassing both renewable and non-renewable sources that adhere to stringent emission standards.
- **Green Energy:** It is a subset of renewable energy that specifically emphasizes on energy sources with the smallest environmental footprint, carbon emissions and other pollutants, focusing on sustainability and ecological preservation.

2. Technological Diversity

- **Renewable Energy:** Focuses on a wide range of technologies such as solar photovoltaic, wind turbines, hydroelectric dams, biomass, and geothermal power.
- **Clean Energy:** Includes various technological solutions beyond renewable sources, such as nuclear energy and carbon capture and storage (CCS).
- **Green Energy:** Concentrates on specific renewable energy technologies known for their minimal environmental impact, such as solar, wind, and hydroelectric power.

3. Policy and Regulation

- **Renewable Energy:** Policies and incentives are mainly aimed at promoting its deployment and reducing reliance on fossil fuels, by setting renewable energy production units.
- **Clean Energy:** Policies are targeted at emissions reduction and environmental protection, including regulations on air quality and subsidies for low-emission technologies.
- **Green Energy:** Supported by policies and initiatives focused on sustainability and environmental stewardship, such as renewable portfolio standards and green procurement programs, aimed at promoting the adoption of environmentally friendly energy sources.

Renewable Energy - "Need of the Hour"

Transitioning to Renewable energy has emerged as the need of the hour due to its pivotal role in mitigating climate change and fostering global sustainability, by significantly reducing carbon emissions compared to fossil fuels. Moreover, renewable energy reduces dependency on finite and geopolitically sensitive resources like oil and gas, thereby addressing geo-political instability and enhancing energy security.

The shift towards renewable energy aligns with sustainability goals, promoting a cleaner, greener, and more resilient energy infrastructure for current and future generations. It also drives economic growth and social development through job creation and technological innovation. As nations worldwide recognize the urgency of addressing climate change, investing in renewable energy stands as a crucial step towards building a more sustainable and prosperous future *(International Renewable Energy Agency – IRENA, 2020).*

Advantages and Disadvantages of Renewable Energy

S.No.	Advantages	Disadvantages
1	**Low Greenhouse Gas Emissions:** Helping mitigate climate change.	**Intermittent Availability:** Renewable energy sources like solar and wind energy depend on weather conditions, leading to fluctuations in power generation.
2	**Energy Security:** Reduces dependence on imported fossil fuels and reducing geopolitical tensions	**Land Use Impact:** It requires significant land use, potentially leading to habitat disruption and land conflicts.
3	**Job Creation:** In manufacturing, installation, and maintenance, fostering economic growth	**Initial Investment Costs:** Can be high, requiring substantial upfront capital, while having lower operating costs over time

S.No.	Advantages	Disadvantages
4	**Sustainable Resource:** Sunlight, wind, and water are abundant and inexhaustible, ensuring long-term energy sustainability.	**Environmental Impact:** Hydropower dams and Biomass plants, may have adverse environmental impacts, including habitat destruction and water pollution
5	**Diverse Applications:** Various utilities, including electricity generation, heating, transportation, and water desalination.	**Grid Integration Challenges:** Integrating into existing grids pose technical challenges, requiring upgrades and investments in grid infrastructure.
6	**Public Health Benefits:** Reduced air pollution from fossil fuel combustion improves public health by decreasing respiratory illnesses.	**Resource Limitations:** Availability of renewable resources may vary regionally, limiting their widespread deployment in certain areas without suitable conditions
7	**Rural Development:** Rural communities are benefited through job creation and land lease payments.	**Energy Storage Challenges:** Storage technologies are still developing and may face cost and efficiency challenges.
8	**Mitigation of Energy Poverty:** It can provide access to clean and affordable energy in remote or underserved regions.	**Regulatory and Policy Uncertainty:** Changing government policies, subsidies, and regulations can affect investment attractiveness of renewable energy projects

Scenario of Renewable Energy in India

India ranks **3rd** in energy consumption and stands **4th** globally in terms of renewable energy installed capacity. According to the Renewable Energy Policy Network for the 21st Century (REN21) report of 2022, India's total installed capacity has reached **179.3 GW** as of July 2023, constituting **43%** of the nation's total capacity. Impressively, this capacity has surged by **396%** over the past 8.5 years, reflecting the rapid pace of renewable energy development in the country. Moreover, India has set an ambitious target of achieving 450 GW of non-fossil fuel-based energy by 2030, marking the world's largest expansion plan in renewable energy. However, challenges such as grid integration, land availability, and financing remain. Continued policy support, technological innovation, and investment will be essential to overcome these challenges and further accelerate the growth of renewable energy in India.

The installed capacity of renewable energy in India is segmented as follows:

- Solar power : 67.07 GW
- Large hydro power : 46.85 GW
- Wind power : 42.8 GW
- Bioenergy : 10.2 GW
- Small hydro power : 4.94 GW

Use of Renewable Energy in Agriculture and Livestock Farming

Electricity stands as a significant yet indispensable expense for farms, exerting pressure on farmers due to escalating costs and energy security uncertainties. However, amidst these challenges, genuine opportunities emerge for farmers to exert greater control over their energy expenditures, anticipate potential disruptions in power supply, and concurrently mitigate greenhouse gas emissions. The percentage share of energy inputs in these sectors varies significantly depending on factors such as farming practices, agricultural land holding, herd size, technological advancements, and regional differences.

In agriculture, energy inputs encompass a range of activities including land preparation, planting, irrigation, fertilization, pest control, harvesting, and transportation. The percentage share of energy inputs in agriculture varies globally, with studies indicating that it can range from 20% to 40% of total recurring costs *(FAO, 2012)*.

Similarly, Power plays a crucial role in various aspects of livestock and poultry farming operations. It is utilized for artificial lighting, to run exhaust fans and heaters. Power is also essential for operating feed mills, grass chaff cutters, and milking machines, feed preparation milk or egg cooling and storage, milk processing, product packaging, and running hatcheries, storage, and distribution of dairy and poultry products.

The energy consumption requirements for various production systems are as follows:

- Cow milk : 2.1 to 5.3 MJ/kg
- Pork meat : 15.9 to 22.7 MJ/kg
- Broiler meat : 9.6 to 19.1 MJ/kg
- Chicken egg production : 20.5 to 23.5 MJ/kg

(Bas Paris et al., 2022)

Different Sources of Renewable Energy

- Biogas
- Solar Energy
- Ethanol
- Biodiesel
- Wind energy

Biogas – "Future of Livestock Industry"

Biogas is produced from organic materials (plant and animal products) by bacteria in an oxygen-free environment, through a process called anaerobic digestion. Biogas systems use anaerobic digestion to recycle these organic materials, turning them into biogas, which contains both energy (gas), and valuable soil products (liquids and solids).

Biogas is typically composed of:

- Methane (CH_4) :50-75%
- Carbon Dioxide (CO_2) :25-50%
- Nitrogen (N_2) :0-10%
- Hydrogen(H_2) :0-1%
- Hydrogen Sulphide (H_2S) :0-3%
- Oxygen (O_2) :0%

(Fact Sheet, EESI, USA, 2017)

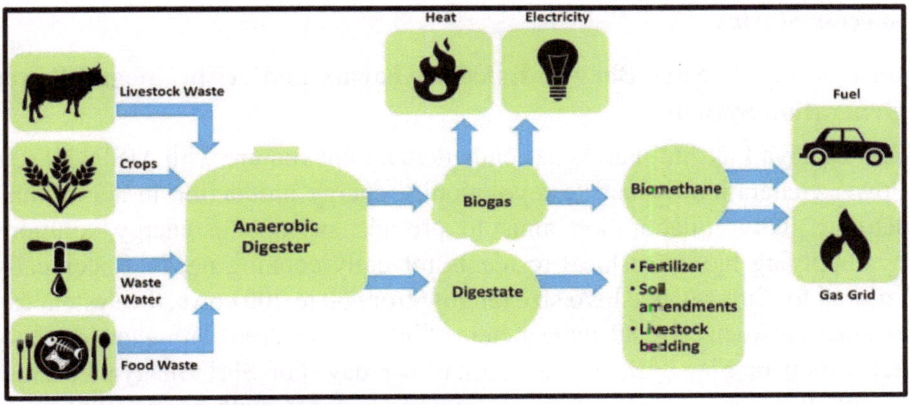

Fig. Illustration of a Biogas production assembly

Mechanism of Biogas production

The mechanism involves three key stages: hydrolysis, fermentation, and methanogenesis.

1. **Hydrolysis:** The complex organic compounds such as carbohydrates, lipids, and proteins are broken down into simpler molecules like sugars, fatty acids, and amino acids, through enzymatic reactions. These hydrolytic enzymes are secreted by microbes.

2. **Fermentation:** During fermentation, anaerobic microbes metabolize the simpler molecules into volatile fatty acids (like acetic acid, propionic acid, and butyric acid), alcohols and other organic acids along with carbon dioxide and hydrogen gas as byproducts.

3. **Methanogenesis:** Methanogenic archaea convert the intermediate products of fermentation into methane (CH_4) and carbon dioxide (CO_2) through anaerobic metabolism. Methanogens only thrive in oxygen-free environments and utilize acetate, hydrogen, and carbon dioxide as substrates to produce methane which is the primary component of biogas.

Bio-Energy – Role in Electrification

Bioenergy refers to energy derived from organic materials or biomass, such as plants, crop residues, animal waste, and organic municipal waste, through various conversion processes. These processes include combustion, fermentation, and chemical reactions, which release energy in the form of heat, electricity, or biofuels. Bioenergy is considered renewable because biomass can be replenished through natural processes and sustainable management practices *(International Energy Agency - IEA, 2021).*

Success Stories

Case Study 1: Shri Bhagya Lakshmi Farms and Aisin Biogas Power Generation System

Shri Bhagya Lakshmi Farms has embarked a joint venture with AISIN Biogas Power Generation System to revolutionize energy production in agricultural settings. This collaboration aims to provide sustainable energy solutions by supplying biogas to local residents for daily cooking needs. Specifically tailored for farms with a herd size ranging from 60 to 200 cows, the system can produce between 25 to 80 cubic meters of biogas daily, translating to electricity generation ranging from 37 to 120 kW per day. For Shri Bhagya Lakshmi Farms, this venture translates into tangible benefits, with every 1000 kg of cow dung input yielding 54 kW of electricity and 4 cubic meters of cooking gas (equivalent to 2 kg of LPG). Remarkably, each kilogram of dung yields 40 liters of gas, underscoring the efficiency of the system. Furthermore, with 1 cubic meter of biogas converting to 6.7 kWh of energy, this collaboration not only enhances energy self-sufficiency but also contributes significantly to environmental sustainability *(Discover Agriculture, March 2021).*

Case Study 2: First Portable Biogas Plant

Mr. Rajagopalan Nair from Thrissur, Kerala manufactured the first portable Biogas plant in 2008. With the help of his Biogas plant, he has eliminated wet wastage from households like discarded meat, chicken, fish-waste, animal dung and vegetable waste by converting them into byproducts like cooking gas and liquid manure. Following his efforts this biogas plant received ISO certification on May 22, 2018 *(Hindustan Times, June 14, 2018).*

Advantages and Disadvantages of Biogas

Advantages	Disadvantages
Renewable Energy Source: Biogas is produced from organic waste, making it a sustainable and renewable energy option *(Weiland, 2010).*	**Little progress in Technology:** Biogas technology may still require advancements to optimize efficiency and reliability *(Kothari et al., 2010).*
Reduces Greenhouse Effect: Biogas production reduces methane emissions from organic waste decomposition, mitigating the greenhouse effect and climate change *(Raheman et al., 2016).*	**Difficult to Increase Efficiency:** Enhancing the efficiency of biogas production processes can be challenging due to factors such as substrate composition and process optimization.
Non-Polluting: Its combustion produces minimal air pollutants compared to fossil fuels, contributing to improved air quality and public health	**Unstable and Flammable:** Biogas is highly flammable, posing safety risks.
Job Opportunities: In construction, operation, maintenance, and management, contributing to local economic development *(Bailis et al., 2015).*	**Contains Impurities:** Biogas may contain impuritie which require purification processes for safe use *(Chin et al., 2015).*

Solar Energy

Solar energy has emerged as a transformative solution for enhancing sustainability and reducing operational costs in agriculture and livestock farming. The installation of solar panels provides a cost-effective means of onsite electricity generation, leading to significant financial savings for farmers *(Shah et al., 2019).* With solar power being abundant and freely available, coupled with a dramatic decrease in the price of solar panels, the adoption of solar energy has become increasingly attractive and feasible for agricultural operations. Moreover, solar panels require minimal maintenance and have a lifespan of up to 25 years, ensuring long-term reliability and efficiency *(Aboumahboub et al., 2020).*

Few solar powered agricultural tools used by farmers

Solar Milking Machine - The Solar panel module is connected to batteries which store electrical energy and power the mobile milking machine. As of now the cost of such machines starts from ₹ 70000. Various state Governments have taken an initiative by providing subsidies on the purchase of such machines leading the way is Karnataka with a subsidy of 50%.

Solar Powered Tractors- They can easily handle planting and harvesting of crops. Sonalika Motors was the first to launch India's first field-ready electric tractor in December 2020 called "Sonalika Tiger Electric".

Solar Water Pumping System- Farmers need to consult the dealers and determine which setup will serve them better based on daily water requirement & geographic location. A single solar pumping system of 1,000Wp can save up to ₹45,000 compared to diesel gensets annually.

Solar Electric Fence- Highly effective and dependable. Consist of SPC, an emitter that produces high voltage impulses (8kV) emitted at intervals of 0.9 to 1.2 seconds and 12V battery. Impulse – 10mA, lasting for a fraction of second. Battery operated solar fences - ₹ 10,000 to 50,000/acre.

(Jayaraman, 2020)

Advantages and Disadvantages of using Solar Energy

Advantages of Solar Energy	Disadvantages of Solar Energy
Most livestock farming regions receive higher sunlight to power a solar photovoltaic (SPV) system *(Kumar et al., 2020).*	High initial cost: The upfront investment required for installing solar panels and associated equipment can be substantial *(Shah et al., 2019).*
After installation, there are virtually no running costs associated with solar energy, leading to long-term financial savings *(Aboumahboub et al., 2020).*	Lengthy payback period: Despite the long-term cost savings, the payback period for solar energy systems can be lengthy, depending on factors such as system size and energy usage *(Bir et al., 2021).*
Solar energy systems are easily expandable, allowing farmers to scale up their energy production to meet growing demands *(Bir et al., 2021).*	Outputs are variable: Solar energy production is dependent on factors such as weather conditions and time of day, leading to variability in energy output *(Shah et al., 2019).*
Solar energy systems can serve as an additional source of income for farmers, as excess electricity generated can be sent back to the grid for compensation *(Gupta et al., 2019).*	

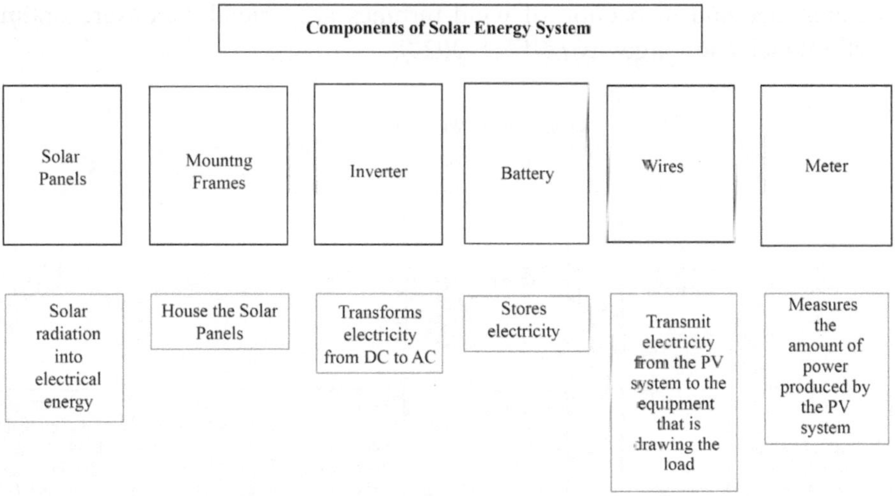

Fig. Various components of solar energy generation system

Biodiesel

Biodiesel, a renewable fuel, is produced through Trans-esterification of vegetable oils, used cooking oils, or animal fats. Biodiesel production in the country primarily relies on imported palm stearin oil. In July 2017, GOI permitted the sale of biodiesel to all consumers for blending with diesel. In year 2019-20 India produced 105.6 million Liters of biodiesel *(Singh et al., 2021).*

Ethanol

Ethanol, a renewable fuel, is produced through the fermentation of sugars found in various grains such as corn, sorghum, barley, sugar cane, and sugar beets. The Indian government introduced the Ethanol Blended Petrol (EBP) Programme nationwide in 2014 to promote the use of ethanol as a blended fuel. As of 2022, ethanol is blended at a rate of 10% in petrol, contributing to cleaner energy and reduced greenhouse gas emissions *(Government of India, Ministry of Petroleum & Natural Gas).*

Wind Energy

Harnessing wind energy through wind turbines presents a promising opportunity for farmers to generate clean electricity on their land. Wind turbines efficiently convert wind energy into electricity, providing a renewable and sustainable power source. However, wind energy is the most variable among all renewable resources, highlighting the importance of conducting a site-specific survey to confirm the reliability of wind resources for a particular location. Regular maintenance and inspection of wind turbines are crucial, to ensure optimal performance and longevity *(AWEA, 2022)*.

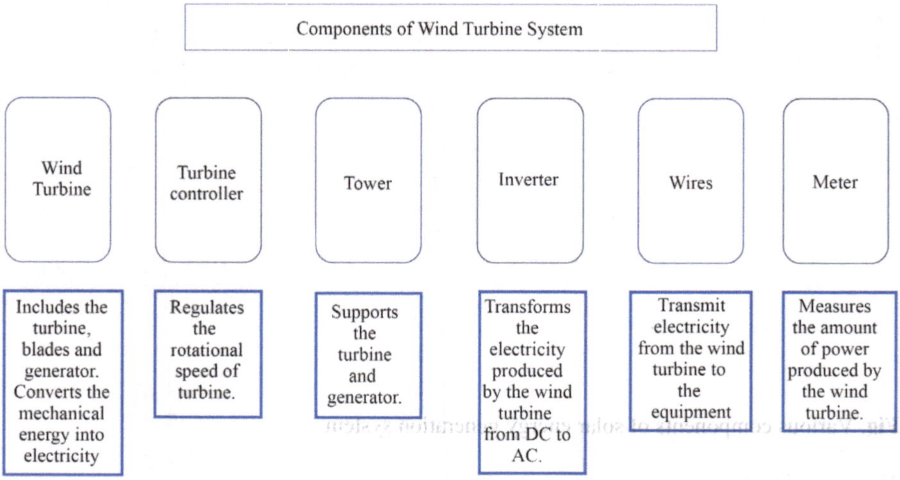

Advantages and disadvantages of wind energy

Advantages	Disadvantages
Wind is a free resource, making wind energy a cost-effective option for electricity generation.	Wind speeds in some farming regions may be inadequate for consistent energy generation, limiting the feasibility of wind power.
Minimal running costs after installation, as wind turbines require little maintenance and no fuel *(Archer & Jacobson, 2005)*.	Output varies based on wind speed, leading to variability in energy production *(Manwell et al., 2009)*.
Wind energy has a low land use footprint compared to other energy sources, allowing for agricultural activities to coexist *(Manwell et al., 2009)*.	Peak power generation from wind energy may not always align with peak electricity demand, requiring grid integration solutions *(Archer & Jacobson, 2005)*.
Wind energy systems can be expanded by adding more turbines, allowing for scalability to meet growing energy needs.	High initial costs and a lengthy payback period may pose financial barriers to wind energy adoption *(Manwell et al., 2009)*.

Government Initiatives towards Renewable Energy

MNRE- Ministry of new and renewable energy

- Earlier known as Ministry of Non-Conventional Energy Sources (MNES).
- Broad aim – To develop and deploy new & renewable energy to supplement the energy requirements of the country.

Objectives of MNRE

1. Energy Security
2. Increase in share of renewable energy.
3. Energy Availability and access
4. Energy Affordability

Schemes – GOI for Promoting Renewable Energy

• NNBOMP (New National Biogas & Manure Programme)

- The program was launched in 2018 aiming to provide clean cooking gas for meeting power and thermal needs of small dairy farmers.
- Subsidy up to Rs 25000 on the establishment of small household biogas plants.

• BPGTP (Biogas Power Generation & Thermal Energy Application Programme)

- The program was started in 2019 to promote biogas based renewable energy sources.
- A 50% subsidy is available for general categories while the program offers 100% subsidy for people belonging to SC & ST categories.

• Waste to Energy Scheme (WTE)

- The scheme was a government initiative to promote setting up of projects for Bio-CNG. It was launched in 2021.
- The scheme offers a CFA up to Rs. 5 lakhs to the farmers.

Future Prospects of Renewable Energy in livestock and Poultry Farming

- Development of new energy future requires research, development, demonstration, deployment and commercialization of new technologies.
- Above mentioned activities must function as part of continuous flow from the research bench to commercial application, with feedback loops.
- Full-fledged Institute on livestock and poultry Renewable Energy Research.

Conclusion

The integration of renewable energy in agriculture, livestock, and poultry farming culminates in a transformative shift towards sustainability and resilience in the agricultural sector. This synergy between renewable energy and agricultural practices offers multifaceted benefits that extend beyond mere energy production. Firstly, the adoption of renewable energy sources such as solar, wind, and biogas reduces dependency on fossil fuels, mitigating greenhouse gas emissions and combating climate change. This not only aligns with global sustainability goals but also enhances the environmental stewardship of farming operations. Secondly, integrating renewable energy technologies enhances the energy security of agricultural enterprises, providing a reliable and decentralized power supply. This resilience is particularly crucial in rural areas where grid connectivity may be unreliable or nonexistent, ensuring uninterrupted operations and safeguarding against energy price volatility. Moreover, renewable energy integration fosters economic viability by reducing energy costs over the long term and diversifying revenue streams through energy surplus sales or incentives such as feed-in tariffs. This financial stability enhances the competitiveness and profitability of agricultural businesses, facilitating long-term growth and investment in sustainable practices. In conclusion, the integration of renewable energy in agriculture, livestock, and poultry farming not only addresses environmental challenges but also fosters resilience, energy security, and economic prosperity. Embracing this transition is imperative for the sustainable development of the agricultural sector, ensuring its viability for generations to come.

References

(National Research Development Corporation, India) https://www.nrdc.org/stories/renewable-energy-clean-facts#sec-whatis (Centre for Resource Solutions, California https://resource-solutions.org/why-renewableenergy/#:~:text=It%20is%20possible%20to%20make,without%20depleting%20the%20Earth's%20resources.

Aboumahboub, T., et al. 2020. "Analysis and comparison of various renewable energy sources and their potential applications in the MENA region." Energy Reports, 6: 1742-1756.

Archer, C. L., & Jacobson, M. Z. 2005. "Evaluation of global wind power." Journal of Geophysical Research: Atmospheres, 110 (D12).

AWEA. (2022). "Wind Energy 101." https://www.awea.org/wind-101#:~:text=Wind%20energy%20(or%20wind%20power,%2C%20pumping%20water%2C%20and%20more

Bailis, R., et al. 2015. "Understanding the job creation potential of resource efficient cities: A case study of bioenergy in the United Kingdom." Energy Policy, 78: 54-63.

Bir, A., et al. 2021. "Feasibility of solar energy in agriculture: A review." Renewable Energy, 175: 1039-1048.

Bull, S. 2001. "Renewable Energy: Ready to Meet Its Promise?" National Renewable Energy Laboratory. https://www.nrel.gov/docs/legosti/old/25890.pdf

Chin, H. S., et al. 2015. "Biogas purification using physical and chemical adsorption methods." Renewable and Sustainable Energy Reviews, 42: 1273-1285.

COP26, United Nations Framework Convention on Climate Change (UNFCCC). 2021. "India's Climate Targets." https://www.climatechangenews.com/2021/11/10/india-announces-

renewable-energy-targets-co2-price-cop26/
Environment and Energy Study Institute (EESI). 2017. "Fact Sheet."www.eesi.org/papers/view/fact-sheet-plug-in-electric-vehicles-2017

Food and Agriculture Organization of the United Nations (FAO). 2012. "The Energy-Water-Food Nexus: A New Approach in Support of Food Security and Sustainable Agriculture." http://www.fao.org/3/i3081e/i3081e.pdf

Government of India, Ministry of Petroleum & Natural Gas. "Ethanol Blended Petrol Programme."https://petroleum.nic.in/sites/default/files/bp%2025.06.2019%20draft%20 report.

Gupta, A., et al. 2019. "Sustainability assessment of renewable energy systems for agricultural production: A review." Renewable and Sustainable Energy Reviews, 113: 109263.

Indian biogas association. 2021. "Biogas to Electricity." Biogas Magazine. Edition 20(12). https://issuu.com/biogas-india/docs/biogas_magazine_edition_20_final/s/16252993

International Energy Agency (IEA). 2021. "Bioenergy." https://www.iea.org/topics/bioenergy

International Renewable Energy Agency (IRENA). 2020. "Renewable Energy and Climate Change." https://www.irena.org/climatechange

Invest India, Government of India (GOI). "Renewable Energy Sector in India." https://www.investindia.gov.in/sector/renewable-energy.

Jayaraman, T. 2020. "Solar Powered Agricultural Tools in India." Ecoideaz. https://www.ecoideaz.com/showcase/solar-powered-agricultural-tools-in-india

Kothari, R., et al. 2010. "Critical review of on-farm anaerobic digestion systems." Bio-resource Technology, 101(19): 8786-8800.

Kumar, A., et al. 2020. "Solar energy potential and its utilization in Indian agriculture: A review." Energy Reports, 6: 1986-1998.

Manwell, J. F., et al. 2009. "Wind energy explained: Theory, design, and application." John Wiley & Sons.

National Renewable Energy Laboratory (NREL). "Advantages and Disadvantages of Renewable Energy Sources." https://www.nrel.gov/research/re-newable-advantages.html

Paris, B., et.al. 2022. "Energy use in open-field agriculture in the EU: A critical review recommending energy efficiency measures and renewable energy sources adoption," Renewable and Sustainable Energy Reviews, Elsevier, vol. 158(C).

Parmar, Ram. 2018. "Ex-Navy man develops portable biogas plant in Vasai." Hindustan Times. https://www.hindustantimes.com/mumbai-news/ex-navy-man-develops-portable-biogas-plant-in-vasai/story-omSdTn4i1SxTIeC0P1XjXL.html

Raheman, H., et al. 2016. "Greenhouse gas emissions from the anaerobic digestion of palm oil mill effluent." Renewable Energy, 96: 1129-1135.

Shah, T., et al. 2019. "Solar energy adoption in agriculture: A review of applications, challenges, and opportunities in global and Indian context." Renewable and Sustainable Energy Reviews, 101: 400-414.

Singh, S., et al. 2021. "Biodiesel as an Alternative Fuel: A Review." Journal of Energy Resources Technology, 143(3): 030801.

U.S. Department of Energy (DOE). "Benefits of Renewable Energy Use." https://www.energy.gov/eere/renewables/benefits-renewable-energy-use

United Nations Framework Convention on Climate Change (UNFCCC). 2021. "Renewable Energy." https://unfccc.int/topics/mitigation/workstreams/renewable-energy.

Weiland, P. 2010. "Biogas production: current state and perspectives." Applied Microbiology and Biotechnology, 85(4): 849-860.

10

Technological Innovations for Holistic Livestock Management

Pankaj Kumar[1], Paradhe Onkar Chandoba[2], Tahir Manzoor[3] Sandeep Bissu[4] and Devkishan[5]

[1]Department of Veterinary Physiology, College of Veterinary and Animal Science, Bikaner (RAJUVAS)-334001, Rajasthan
[2]Department of Animal Nutrition, College of Veterinary and Animal Science, Bikaner (RAJUVAS)-334001, Rajasthan
[3]Department of Livestock Production Management, College of Veterinary and Animal Science, Bikaner (RAJUVAS), Rajasthan
[4]Department of Veterinary Physiology, College of Veterinary and Animal Science, Bikaner (RAJUVAS), Rajasthan
[5]Department of Veterinary Pharmacology and Toxicology, College of Veterinary and Animal Science, Bikaner (RAJUVAS), Rajasthan

Abstract

The livestock systems that gave rise to the third agricultural "green" revolution in the middle of the 20th century were constantly undergoing technological modification. We are currently entering the fourth agricultural revolution, driven by technological advancements such as digital technology. In this revolution, data collection technologies that increase farm and supply chain performance, task automation, and compliance will assist the production of sustainable food. The efficiency and profitability of animal production increase with the use of technological instruments. Thus, the primary field of study for animal productivity and sustainability involves technological developments. Leveraging innovation and technology is essential to realise the full potential of holistic livestock management. The possibilities for boosting production, efficiency, and sustainability are numerous, ranging from robots and automation to data-driven decision-making and precision animal husbandry. Livestock farmers who adopt these innovations will be better equipped to deal with the demands of a changing global landscape, improve animal welfare, make the most use of available resources, support more sustainable agriculture, and provide holistic livestock management.

Using cutting-edge technology, precision livestock farming maximises the health, welfare, and productivity of animals. Technological advancement may carefully monitor and manage individual animals or groups with the use of tools like automated systems, smart ear tags, and remote monitoring. This chapter provides insightful information on the most recent significant technologies that enhance holistic livestock management.

Keywords: *Revolution, Innovation, Technology, Sustainability, Management, etc.*

1. Introduction

Technology innovation is characterised by the development and utilisation of novel or enhanced technologies, instruments, frameworks, and procedures that result in noteworthy progressions or discoveries across diverse domains. It entails utilising resources, knowledge, and experience to create creative solutions that address issues, boost productivity, advance the cause, and add value.

In numerous crucial areas, the significance of technological innovation is evident:

- Improved Quality of Life
- Economic Growth and Competitiveness
- Enhanced Efficiency and Productivity
- Addressing Societal Challenges
- Scientific and Technological Advancement
- User Empowerment and Engagement
- Sustainable Development

1.1. Importance of new technology

With the aid of modern technology, farmers may increase productivity, profitability, and the care of their cattle while working more easily. With the use of management and direct manipulation capabilities, technological advancements provide farmers with faster, more profitable, and more effective ways to complete tasks on schedule. For the welfare of animal management, the disease must be carefully managed and continuously monitored. This can be accomplished by identifying the infection in its early stages and then identifying and treating it. Modern automation consists of machinery along with extremely advanced software and electronics. For easier animal husbandry, several computer-assisted picture analysis tools are being developed. The newest computer programs can recognise and categorise animal sounds according to certain circumstances.

1.2. Benefits of new technologies

- Cost efficiency increases
- Improvement in animal welfare
- Working conditions are improved
- Better production monitoring such as remote monitoring, access to real-time data, etc.
- Improved provision of important production data

1.3. Need for newer technologies in the following areas for holistic livestock management

- Electronic records
- Milking
- Heat detection
- Walk-over-weighing
- Auto-drafting
- Genetic improvement
- Feeding
- Barn-environment optimisation
- Health-recording, etc.

NOTE: In animal husbandry, appropriate objective measuring systems are required to promptly and safely identify disease, a typical oestrus cycle, silent heat, and stress in animals.

2. Automation

2.1. Technological advancements in milking automation

a) **Automatic milking systems:** To protect the quality of the milk and the health of the animals, an automated milking system necessitates an entirely different management system for feeding, milking, cow traffic, cow behaviour, and grazing.

Use: - Information related to milk production, speed of milking, milk acidity, milk conductivity, etc. can be obtained.

b) **Electronic devices or Sensors:** To detect anomalies, human visual inspection must be replaced with electronic devices or sensors. A definition of abnormal milk remains one of the fundamental prerequisites for the development of sensors to detect it.

   Use: - Milk progesterone level, the temperature of milk, etc. can be obtained.

c) **Milking robots:** Fitted with sensors, which assess several characteristics of unusual milk pH, somatic cell count, acidity, conductivity, etc., aid in the detection of mastitis symptoms.

d) **Simple automatic cup removal device:** It keeps track of each cow's milk flow rate. When they reach a certain level, the milking vacuum is turned off, and the system is triggered to remove the cups from the cow.

e) **Post-milking teat disinfection:** One well-established element of many mastitis control plans is post-milking teat cleaning. Many farmers often carry out this manually using a dip cup or a pressure-operated spray lance.

f) **Behaviour meter:** For animal monitoring, behaviour meters are also installed in the milking systems. The behaviour meter keeps track of each animal's activity level, lying time, and lying sessions continuously. The ability to analyse the reproductive and health state of animals, as well as their welfare in various environmental circumstances and stressful situations, is made possible by the observations of animal behaviour.

g) **Separation gate:** Used in automatic management systems.

Note: Newly developed sensors such as NIR (Near Infra-Red) are used in automatic milking systems yield far faster and more efficient results.

2.2. Technological advancements in feeding automation

It consists of whole systems, which include installations for feed distribution, mixing equipment, feed preparation, and all feeding stages. Feed components will be loaded, mixed, and provided to the feed table that the systems have built up, including grass and maize silage, mineral feed, and feed concentrate. A control panel, a programmed command manager, a scale, a communication interface, and all the equipment required to arrange feeding and provide food to animals in each age group, make up the automation systems.

a) **Computer-controlled calf feeders:** Compared to conventional calf feeding techniques, computer-controlled calf feeders offer numerous benefits. Calves are equipped with transponders, which allow one to monitor each calf's daily consumption. The computerised milk feeding technique is quite easy for calves to learn, and it saves a lot of labour costs—roughly 73%. For the well-being of the calves, these systems can be paired with automated weighing and health monitoring systems.

b) **Electronic Concentrate Feeding system:** With the help of this method, every cow is guaranteed to receive the precise amount of feed at the precise moment.

c) Sensors

i. **Rumen activity sensors:** Rumen activity sensors are one of the most well-liked and cutting-edge methods used by cattle producers to lower metabolic disorders. When the sensitive animal shows signs of increasing acidosis, a farmer can modify their nutrition to avoid more serious issues.

ii. **Electronic sensors:**

- The temperature and pH of cattle rumen can be measured using a variety of electronic sensors. In particular, rumen bolus may function continuously for 100 days, and data is saved every 15 minutes for analysis at a later time. Rumination activity serves as a reliable gauge of the health of cattle. Rumination requires a certain degree of well-being; rumination is inhibited by anxiety and other illnesses.
- Another sensor is used to measure the amount of chewing action by gathering data on cow jaw movement. This sensor operates on the notion that when an animal opens and closes its mouth, its changing pressure is not sensed.

3. **Electronic identification**

3.1. Radio Frequency Identification (RFID) system

- Electronic ID system started in the 1970s.
- Unique electronic number for an individual animal. This number links that animal to the database.
- Work using radiofrequency for sending data, generally placed subcutaneously.

Types of RFID tags

a) **Boluses with a transponder:** Boluses are recognised as safe for the health of animals after they pass through the ruminants' first two stomachs. With a 100% retention rate, they can even be given to lambs who have been weaned at the fifth week.

b) **Injectable glass tags/injectable transponders:** These can be applied easily after birth, while the preferable locations differ in each animal species.

c) **Ear tags**

Functioning of RFID tags

- All these devices have a special chip system for sending data to the base computer for evaluation.
- These devices have some specific components on their system regarding storing and evaluating data used for evaluating herd data.
- Some electronic tags have a reader that can receive and store the required data for evaluation.
- Some of the tag works transferring the number to another storage system for another evaluation stage.
- Data is sent using an antenna to transfer data on the system.

c.2. Electronic weighing system: A robust and user-friendly method for precisely weighing animals. Hence, producers may simply and consistently check the performance of their animals. Data that has been stored is sent to the main computer for analysis.

c.3. Auto-drafting: Cattle going through a race are automatically separated based on age, sex, weight, or any other criteria the farmer prefers.

4. Artificial intelligence (AI)

4.1. Livestock Management Through Artificial Intelligence-based Technology: In modern animal husbandry, animals of different weights and stages can be identified by using artificial intelligence. This can provide them with a variety of diets and improve the production rate of high-quality animals. Artificial intelligence is currently being used for animal monitoring, disease prediction, and disease diagnosis. These technologies reduce repetitive tasks in livestock farms and improve animal husbandry, ensuring more profits for the business. A.I. technology has been successfully adopted by many industries.

Now it is set to revolutionize the future of animal husbandry with drones, robots, and intelligent monitoring systems.

4.2. Smart Animal Enclosure: Habitat management of animals is being done through artificial intelligence-based technology. Nowadays, in the smart livestock sector, animals are being followed for accurate monitoring of health. For this, cameras and artificial intelligence (AI) are used. Through this technology, injuries and diseases in animals can be detected quickly through detailed observation and image-based analysis. This can improve the quantity and quality of milk production.

4.3. Computer Vision for Animal Monitoring: Monitoring animals through sensors is common. These require the use of physical tracking devices on a large scale. This allows individual monitoring of group behaviour, early detection of lameness, and recording of eating habits. With this, the animal's loss of appetite, discharge from the eyes or nose, etc. can be detected quickly.

4.4. Other Uses of Artificial Intelligence in the Livestock Sector: AI in the livestock sector There are many other applications like early detection of injuries, fertility monitoring, anti-stress tags for cattle, virtual fence for cattle, poaching prevention, large-scale grazing of large cattle herds, cattle tracker, automated milking, milking robot, cow gait analyser, etc. can be mentioned.

5. Precision livestock farming (PLF)

With the use of process engineering principles, precision livestock farming (PLF) technologies automate livestock agriculture and enable farmers to keep an eye on large animal populations for health and welfare, identify problems with individual animals quickly, and even predict problems based on historical data.

Advancements in PLF technologies

- Tracking the behaviour of cattle
- Identifying respiratory illnesses in numerous species by monitoring coughs
- Identifying pregnancy in bovines by analysing changes in body temperature
- Detecting vocalisations in pigs, such as screams.

Use of PLF technology

- Assist farmers in keeping an eye out for infectious diseases in livestock agriculture, hence enhancing the availability and safety of food.

- Food safety concerns may be minimised
- Resource efficiency can be maximised
- Animal health and welfare improvement

5. Challenges and Way Forward

Livestock management faces various difficulties that require careful consideration and creative solutions to maintain long-term profitability. These issues cover a wide range of dairy farming-related topics, such as animal welfare and health, reproductive effectiveness, environmental impact, and technical improvements. Numerous research has examined these issues and offered suggestions for the management of livestock in the future.

There are promising ways to lessen the ecological footprint of livestock, including the use of sustainable techniques that include enhanced waste management, precision nutrition management, and the use of renewable energy. Furthermore, it is highly recommended that livestock management adopt creative and sustainable methods. It may be possible to improve ecological responsibility, precision farming methods, and breeding strategies by combining traditional methods with contemporary technology. Key factors for guaranteeing long-term sustainability are emphasizing animal welfare, reducing environmental effects, and increasing output.

The necessity of accurate and timely monitoring of each animal's behaviour, health, and nutritional needs is one of the potential difficulties. To meet this challenge, wearable technology and automated monitoring systems must be developed and seamlessly integrated with other cutting-edge sensor technologies to collect data on feed intake, growth trends, and physiological characteristics. Moreover, the integration of artificial intelligence and machine learning algorithms is assumed to be of utmost importance in the analysis and interpretation of data, which eventually aids in the development of precise and customised feeding strategies.

6. Conclusion

The key to realising livestock farming's full potential is embracing innovation and technology. The possibilities for boosting production, efficiency, and sustainability are numerous, ranging from robots and automation to data-driven decision-making and precision animal husbandry. Livestock farmers may better manage the demands of a changing world, improve animal welfare, make the most use of available resources, and help create a more sustainable agriculture industry by adopting these innovations. Together, we can leverage technology to create a more promising future for sustainable food production and cattle rearing. Numerous studies concluded that these applications might be utilised to track the well-being of animals and offer early disease, physiological conditions, and abnormalities identification.

Although technology has a lot of promise, it also poses difficulties for livestock farmers. There is a need to examine frequent obstacles to technology adoption, such as expense, lack of technical expertise, and restricted access to infrastructure. There should be some solutions to overcome these obstacles, such as cooperation between parties, government assistance, and training initiatives. Social and economic changes that promise to create a digitally inclusive and healthy society through innovations in livestock farming digitalization solutions necessitate citizen participation and involvement through the co-creation of the technology development and its validation.

References

Artmann, R. 1999. Electronic identification systems: state of the art and their further development. Computers and Electronics in Agriculture, 24(1-2): 5-26.

Benjamin, M., & Yik, S. 2019. Precision livestock farming in swine welfare: a review for swine practitioners. Animals, 9(4): 133.

Caja, G., Conill, C., Nehring, R., & Ribo, O. 1999. Development of a ceramic bolus for the permanent electronic identification of sheep, goat and cattle. Computers and Electronics in Agriculture, 24(1-2): 45-63.

Caja, G., Hernandez-Jover, M., Conill, C., Garin, D., Alabern, X., Farriol, B., & Ghirardi, J. 2005. Use of ear tags and injectable transponders for the identification and traceability of pigs from birth to the end of the slaughter line. Journal of Animal Science, 83(9): 2215-2224.

De Koning, C. J. A. M. 2004. Y van de Vorst Meijering A. Automatic milking experience and development in Europe. In Proceedings of the first North American Conference on Robotic Milking, Toronto, Canada (pp. I1-I11).

Eastwood, C. R., Edwards, J. P., & Turner, J. A. 2021. Anticipating alternative trajectories for responsible Agriculture 4.0 innovation in livestock systems. Animal, 15: 100296

Exadaktylos, V., Silva, M., Aerts, J. M., Taylor, C. J., & Berckmans, D. 2008. Original papers: Real-time recognition of sick pig cough sounds. Computers and Electronics in Agriculture, 63(2): 207-214.

Ferrari, S., Piccinini, R., Silva, M., Exadaktylos, V., Berckmans, D., & Guarino, M. 2010. Cough sound description in relation to respiratory diseases in dairy calves. Preventive Veterinary Medicine, 96(3-4): 276-280.

Garin, D., Caja, G., & Bocquier, F. 2003. Effects of small ruminal boluses used for electronic identification of lambs on the growth and development of the reticulorumen. Journal of Animal Science, 81(4): 879-884.

Ghasura, D. 2023. Embracing Technology and Innovation in Livestock Farming: Paving the Path to Sustainable Agriculture. https://www.linkedin.com/pulse/embracing-technology-innovation-livestock-farming-paving-ghasura

Gheorghe-Irimia, R. A., Sonea, C., Tapaloaga, D., Gurau, M. R., Ilie, L. I., & Tapaloaga, P. R. Innovations in Dairy Cattle Management: Enhancing Productivity and Environmental

Sustainability. Annals of' Valahia" University of Târgovişte. Agriculture, 15(2): 18-25.
Goncu, S., & Gungor, C. 2018. The Innovative Techniques in Animal Husbandry. In Animal Husbandry and Nutrition. IntechOpen. https://doi.org/10.5772/intechopen. 72501
Hogeveen, H., Kamphuis, C., Sherlock, R., Jago, J., & Mein, G. 2009. Inline SCC monitoring improves clinical mastitis detection in an automatic milking system. In Precision livestock farming'09 (pp. 315-322). Wageningen Academic.
Jain, N. 2023. What is Technology Innovation? Definition, Examples, and Strategic Management. IdeaScale. https://ideascale.com/blog/what-is-technology-innovation/
Kamphuis, C., Mollenhorst, H., Feelders, A., & Hogeveen, H. 2008. Decision tree induction for detection of clinical mastitis using data from six Dutch dairy herds milking with an automatic milking system. Mastitis Control–From Science to Practice. The Hague, the Netherlands, 267-274.
Kilic, U. 2011. Use of wireless rumen sensors in ruminant nutrition research. Asian Journal of Animal Sciences, 5(1): 46-55.
Lee, C. S., Wooding, F. P., & Kemp, P. 1980. Identification, properties, and differential counts of cell populations using electron microscopy of dry cow's secretions, colostrum and milk from normal cows. Journal of Dairy Research, 47(1): 39-50.
Neethirajan, S., & Kemp, B. 2021. Digital livestock farming. Sensing and Bio-Sensing Research, 32: 100408.
Neethirajan, S., Ragavan, K. V., & Weng, X. 2018. Agro-defense: Biosensors for food from healthy crops and animals. Trends in Food Science & Technology, 73: 25-44.
Neethirajan, S. 2017. Recent advances in wearable sensors for animal health management. Sensing and Bio-Sensing Research, 12: 15-29.
Norton, T., Chen, C., Larsen, M. L. V., & Berckmans, D. 2019. Precision livestock farming: Building 'digital representations' to bring the animals closer to the farmer. Animal, 13(12): 3009-3017.
Owens, F. N., Secrist, D. S., Hill, W. J., & Gill, D. R. 1998. Acidosis in cattle: a review. Journal of animal science, 76(1): 275-286.
Pyorala, S. 2002. New strategies to prevent mastitis. Reproduction in domestic animals, 37(4): 211-216.
Rainard, P., & Riollet, C. 2006. Innate immunity of the bovine mammary gland. Veterinary research, 37(3): 369-400.
Reinemann, D. J., & Helgren, J. M. 2004. Online milk sensing issues for automatic milking. In ASAE/CSAE Annual International Meeting.
Sonea, C., Gheorghe-Irimia, R. A., Tapaloaga, D., Gurau, M. R., Udrea, L., & Tapaloaga, P. R. Optimizing Animal Nutrition and Sustainability Through Precision Feeding: A Mini Review of Emerging Strategies and Technologies. Annals of' Valahia" University of Târgovişte. Agriculture, 15(2): 7-11.
Sordillo, L. M., Shafer-Weaver, K., & DeRosa, D. 1997. Immunobiology of the mammary gland. Journal of dairy science, 80(8): 1851-1865.
Voulodimos, A. S., Patrikakis, C. Z., Sideridis, A. B., Ntafis, V. A., & Xylouri, E. M. 2010. A complete farm management system based on animal identification using RFID technology. Computers and electronics in agriculture, 70(2): 380-388.

Index